Level 5

Science

McGraw, Tondy, Toukonen

Bright Ideas Press, LLC
Cleveland, OH

Simple Solutions Level 5
Science

All rights reserved. No part of this publication may be reproduced or transmitted in any form or by any means, electronic or mechanical, including photocopy, recording, or any information storage or retrieval system. Reproduction of these materials for an entire class, school, or district is prohibited.

Printed in the United States of America

ISBN-13: 978-1-934210-63-5
ISBN-10: 1-934210-63-3

Cover Design: Dan Mazzola
Editor: Kimberly A. Dambrogio
Illustrator: Christopher Backs

Copyright © 2010 by Bright Ideas Press, LLC
Cleveland, Ohio

Welcome

Simple Solutions.
Minutes a Day-Mastery for a Lifetime!

Note to the Student:

    We hope that this program will help you understand Science concepts better than ever. For many of you, it will help you to have a more positive attitude toward learning these topics.

    Using this workbook will give you the opportunity to remember topics you have learned in previous grades. By revisiting these topics each day, you will gain confidence in Science.

    In order for this program to help you, it is extremely important that you do a lesson every day. It is also important that you ask your teacher for help with the items that you don't understand or that you get wrong on your homework.

    We hope that through Simple Solutions and hard work, you discover how satisfying and how much fun Science can be!

Lesson #1

Scientific Inquiry

An inquiry is an **investigation**. It is an organized way to find answers to questions or solutions to problems. A scientist uses the **scientific method** during an inquiry. Everyone – including you – can use the scientific method to investigate a problem or search for the answer to a question. Here are the steps:

- Question
- Research
- Form a Hypothesis
- Experiment
- Gather Data
- Share Conclusions

In the first step, you will choose a problem or something that you are curious about, and then write a **question**.

1. Scientists use the _____ to investigate a problem or search for the answer to a question.

2. Asking a _____ is the fist step of the scientific method.

3. Who uses the scientific method?
    A) scientists
    B) scientists and college professors
    C) students in a science lab
    D) anyone can use it

4. Three of the terms below have something in common. Cross out the one that doesn't belong.

    hurricane   volcano   tsunami   tornado

5. In the item above, what do the three terms have in common?
    A) They name fictional events.
    B) They name severe weather events.
    C) They only occur at night.
    D) They only occur during the daytime.

Simple Solutions© Science   Level 5

6. An animal with a backbone is called a(n) _____.

   invertebrate   exoskeleton   mollusk   vertebrate

7. A mass of slow-moving ice is called a _____.

   delta   mountain   glacier   geode

Photosynthesis

The chlorophyll in plant leaves absorbs sunlight.

Plants take in carbon dioxide from the air.

Plants release oxygen into the air.

Plants store food in their stems, leaves, and seeds.

Plants absorb water and nutrients through their root system.

8. Study the illustration above. What are **three** things that plants use during the process of photosynthesis?

   water   clouds   sunlight   animals   carbon dioxide

9. Where do plants store the food that they make?

   _____

10. What is the gas that plants give off during photosynthesis?

    _____

3

Lesson #2

Scientists use tools to observe and measure things both in a lab and in the field during scientific investigations. Each tool is designed to do a specific job. For example, a **microscope** and a **hand lens** both allow you to *magnify* or make objects appear bigger, so you can see them in greater detail.

1. What tools or equipment will you use during scientific investigations? Match each tool with its use. If you aren't sure, check the Help Pages.

   ___ microscope   A) measures liquid volume

   ___ beaker   B) measures length

   ___ ruler   C) magnifies objects (easy to use outside a lab)

   ___ hand lens   D) magnifies objects (thousands of times bigger)

Scientific Inquiry

The second step of the scientific method is **research**. In this step, you will gather as much information as you can from books, magazines, the internet, and other people. You will use your research to help you do the following: prepare a safe experiment, choose proper tools and procedures, and write a hypothesis.

2. Why is it important to do research as part of the scientific method?
   A) Research will help you to plan a safe experiment.
   B) Research will help you know what tools to use.
   C) Research will help you to form a solid hypothesis.
   D) All of the above are reasons to do research.

3. What is the first step of the scientific method? _____

4. An ice cube is which of these?   solid   liquid   gas

5. Weathering breaks rock into sediments. Water, wind, and moving ice carry away the sediments in a process called _e_ __ __ __ __ __ __ .

Simple Solutions© Science   Level 5

Lab Safety

People who work in labs must also follow safety rules. Here are a few lab safety rules:

Follow your teacher's directions. Chemicals, lab equipment, and sharp objects may be dangerous, and your teacher knows how to use them properly. Be safe in the lab – pay attention and don't ever act silly.

Dress Appropriately – Always wear safety goggles when you are working with any kind of chemical. Your teacher may ask you to wear gloves to protect your hands from chemicals. Wear sturdy shoes – not sandals – and put on a lab coat or apron to protect your clothing. Tie back long hair and remove any dangling jewelry.

6. Why would it be better to wear long pants instead of shorts in a science lab?

   A) Long pants are warmer.
   B) Long pants will cover your legs and protect them.
   C) Short pants are never appropriate for school.
   D) Shorts are not very comfortable.

7. When should you wear goggles? Check all that apply.

   ___ when working with chemicals

   ___ when the teacher tells you to

   ___ while looking into a microscope

   ___ when studying for a science exam

8. The solid rock that forms the Earth's surface is called _____.

   soil   bedrock   mountains   desert

9. Animals like jellyfish, snails, and earthworms do not have backbones. These animals are called _____.

   invertebrates   exoskeletons   mollusks   vertebrates

10. All animals have four **basic needs**. Fill in the missing ones.

    _____   _____

    _____air_____   _____shelter_____

Lesson #3

1. What tools or equipment will you use during scientific investigations? Match each tool with its use. If you aren't sure, check the Help Pages.

    ____ thermometer   A) measures an object's mass

    ____ anemometer   B) measures temperature

    ____ spring scale   C) measures weight or friction

    ____ balance   D) measures wind speed

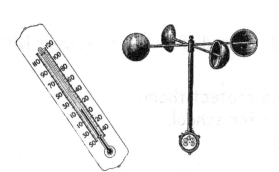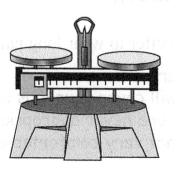

2. Why should you put on a lab coat during an experiment?

    A) A lab coat will protect clothing.
    B) A lab coat looks professional.
    C) A lab coat will keep you warm.
    D) A lab coat has pockets which you need during an investigation.

3. Earth's surface is made of solid rock called <u>b</u> __ __ __ __ __ __ .

Scientific Inquiry

The third step of the scientific method is to formulate a **hypothesis** or "educated guess." You can make an educated guess once you have some information from your research. In this step, you will write a hypothesis as a statement, and make it as specific as possible.

4. Which of these hypotheses is most specific?

    A) My heart rate will be higher after doing jumping jacks for 1 minute.
    B) Exercise is good for a person's heart.
    C) How do jumping jacks affect heart rate?
    D) Exercise will increase your heart rate.

5. Write a hypothesis to go with this question: Will bean plants grow faster in direct or indirect sunlight?

   _____

   _____

Photosynthesis

Plants make their food through **photosynthesis**. To make their food, plants use water, carbon dioxide, and energy from the sun. During photosynthesis, plants make food in the form of sugar, and they give off oxygen which is something all animals need. Oxygen is a by-product of photosynthesis.

6 – 10. Use the word bank to complete the sentences below.

oxygen

water

photosynthesis

sunlight

carbon dioxide

food

During the process of _____, plants take in

_____, _____, and _____.

Plants make _____ and give off _____

as a by-product of photosynthesis.

Simple Solutions© Science   Level 5

Lesson #4

What Happens During Photosynthesis?

1. Fill in the missing words to complete the illustration below. If you need help, look back at Lesson #1.

A) The _____ in plant leaves absorbs sunlight.

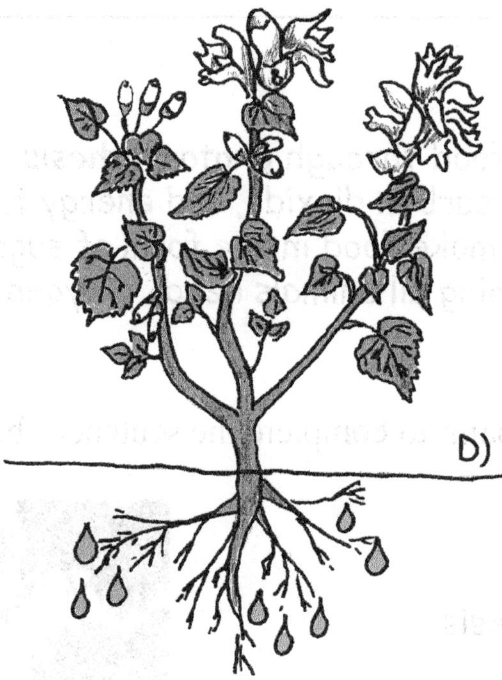

B) Plants release _____ into the air.

C) Plants take in _____ from the air.

D) Plants store food in their _____, _____, and _____.

2. Which **three** things do plants need in order to survive?

   air   light   pesticides   water   shelter

3. Two of the four basic needs of animals are listed below. What are the other two?

   __shelter__   _____

   __oxygen__   _____

8

Simple Solutions© Science   Level 5

Scientific Inquiry

The fourth step of the scientific method is **experiment**. This step includes two things: a **list of materials** and a **procedure** (step by step explanation of what to do). The purpose of an experiment is to <u>test</u> a hypothesis. The experiment may show that your hypothesis is correct. Or, it may show that your hypothesis is incorrect. Or, it may not show anything! If the experiment does not prove the hypothesis was correct, this does not mean the experiment was a failure.

Let's say you hypothesize that plant food will make a plant grow faster. You set up a procedure to <u>test</u> your hypothesis: Get two identical plants, give them the same amount of water, and put them next to each other on the same window sill. Add plant food to one of the plants but not the other. Observe daily and record what you see.

After several weeks, both plants are healthy and exactly the same size. Your experiment is not a failure, but your results are **inconclusive**. That means the experiment did not prove your hypothesis. Scientists may repeat or **replicate** an experiment several times before coming to a solid conclusion.

4. What is the purpose of an experiment? _____

5. What does *inconclusive* mean?

    A) not proving anything
    B) incorrect
    C) step by step explanation
    D) failure

6. What does *replicate* mean?   try   test   repeat   experiment

7. Name two things that are part of an experiment.

    _____   _____

8. Which of these is the best way to describe a hypothesis?
    random thought   educated guess   proven theory   body of research

Read each statement. Write **T** for true or **F** for false.

9. _____ If an experiment does not prove a hypothesis, it means the experiment was faulty.

10. _____ A scientist may repeat an experiment several times.

Lesson #5

What makes a food chain?

Energy travels between **organisms** (living things) in a **food chain**. You know that energy comes from the sun. Plants use the sun's energy to make food, and that is how plants live and grow. Some animals eat plants to get energy, and that is how they live and grow. Other animals get their energy by eating animals that eat plants.

Study the illustration of a food chain; then answer the questions below.

1. Which of the following should be at the bottom of the food chain?

   squirrel   human   shark   grass

2. What do you think would happen if the frogs in this chain became extinct?

   A) There would be more grasshoppers.
   B) There would be fewer snakes.
   C) There would be fewer hawks.
   D) All of these could happen.

3. Animals that eat other animals are called **carnivores**. Which animals in the food chain at the right are carnivores?

   grasshopper   frog   snake   hawk

4. During the winter some animals go into a deep sleep called _____.

   A) migration   C) hibernation
   B) camouflage   D) metamorphosis

_____?_____

10

5. _____ is a force that pulls objects toward Earth.

    Magnetism   Motion   Gravity   Energy

6. Which would you use to examine a grasshopper?

    balance   spring scale   hand lens   dropper

Scientific Inquiry

The next step in the scientific method is to **gather data**. During an experiment, you will gather valuable data (facts, numbers, and other information). It is best to record data while you are performing the experiment, if possible. Take careful notes because it is difficult to remember specific information after time has passed. You will record and organize the data so that it makes sense and is easy to share with others.

Your data may prove or disprove your hypothesis. Or, your results may be inconclusive (uncertain). If your data is inconclusive, you may decide to repeat the experiment or design a new one.

7. Is the following statement true or false?

    _____ An experiment is meant to prove that the hypothesis is correct.

8. When is the best time to gather data about your experiment?

    before   during   after   anytime

9. Which of these can be recycled? Underline the names of any that can.

    A) aluminum cans   C) newspapers
    B) plastic milk bottles   D) Styrofoam™ cups

10. Which of these is a **renewable natural resource**?

    plastic   water   petroleum   coal

Lesson #6

Scientific Inquiry

The final step of the scientific method is to draw a **conclusion**. In this step, you will analyze your data a little more and summarize what you learned. You will share your conclusions with others. You may write a report, give an oral presentation, or display your work as part of a science fair.

Gina had this hypothesis: *More birds will visit my backyard if there is a birdseed feeder in it.* Gina filled a bird feeder with seed and hung it on a tree branch. She checked it each morning for a week. Each morning, almost all of the seed was gone, but Gina did not count very many birds. Finally, she noticed that the birdseed feeder was coming apart on one side. Squirrels were jumping onto the feeder from the tree branches. When the seed spilled out, the squirrels had a feast.

1. As a result of her experiment, what can Gina conclude?

    A) The hypothesis was correct.
    B) The hypothesis was not correct.
    C) The experiment was not a good way to test the hypothesis.
    D) The results were inconclusive.

2. If Gina still wants to test her hypothesis, what should she do next?

    A) Repeat the experiment with a different birdseed feeder.
    B) Take pictures of the squirrels eating the seed.
    C) Repeat the experiment, but use bread crumbs instead of seed.
    D) Repeat the experiment, but change the hypothesis.

3. The purpose of an experiment is to _____ a hypothesis.

Label the three main parts of a plant.

4. A) _____

5. B) _____

6. C) _____

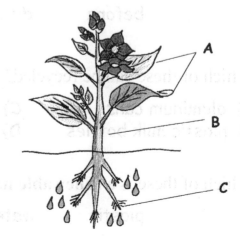

Lab Safety

Here are a few more lab safety rules:

If an experiment involves tasting or smelling something, your teacher will give you special permission. Otherwise, **never eat or drink anything in the lab or during a lab activity** and **never taste or sniff chemicals.** Breathing in chemicals can harm you, and eating chemicals can make you sick. **Wash your hands after handling anything in the lab.** Do not ever use lab equipment like beakers to drink from.

7. Why should you wash your hands after working with lab equipment?
   A) People who touched the lab equipment may be sick.
   B) Harmful chemicals may have gotten onto your hands when you touched the lab equipment.
   C) Lab equipment smells bad.

8. When do you think it would be okay to sniff or taste something during an experiment?
   A) When the teacher says it's okay.
   B) When your friends say it's okay.
   C) When you're very hungry.
   D) Anytime, as long as you are careful.

9. Which of these is an example of a mammal? Check the Help Pages if you're not sure.

   rabbit   snake   alligator   hawk

10. Look at the teeth in each skull below. Two skulls have similar teeth and one is different. Which skull is probably from an animal that ate grass?

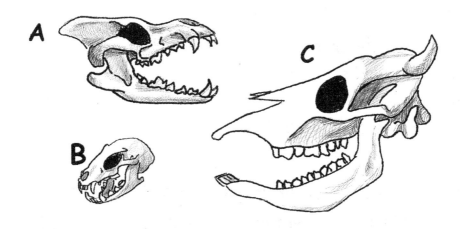

Lesson #7

1. The _____ is a procedure scientists use to find out about the world around us.

    water cycle   scientific method   solar system   cause and effect

2. Which weather event names *a violently rotating column of air which may appear as a funnel cloud*?

    tsunami   blizzard   tornado   earthquake

3. Animals that do not have backbones (like insects, worms, and jellyfish) are called _____.

    invertebrates   vertebrates   carnivores   none of these

4. Which scientific instrument would you use to examine the leaf of a plant?

    hand lens   measuring cup   balance   forceps

In order to produce food, plants must be able to absorb light energy from the sun. **Chlorophyll** in the plant cells makes it possible for the plants to absorb light energy. Chlorophyll is also the pigment (coloring) that makes plants green. In autumn there are fewer hours of daylight, so many plants and trees stop making chlorophyll. The other pigments in their leaves become visible, and that makes autumn very colorful!

5. Plants make their own food through a process called _____.

    _____

6. Which of these is one main function of a plant root?

    A) to anchor the plant in the soil
    B) to make food for the plant
    C) to take in sunlight
    D) to produce carbon dioxide

7. Without _____, there would be no life on Earth.

    plants   water   sunlight   all of these

8. How do the seeds from plants spread and grow in other areas?

   A) Animals carry seeds on their fur and in their digestive systems.
   B) Wind and water carry seeds to new places.
   C) Birds carry seeds as food and drop them along the way.
   D) All of the above are ways that seeds travel.

The Transfer of Energy in the Food Chain

All living things need energy to live and to grow. Plants get energy from the sun. Animals get energy by eating plants or by eating animals that eat plants. This transfer of energy from one living thing to another is called a **food chain**. Usually you won't see the sun as part of a food chain, but the transfer of energy starts with the sun.

Use words from the example to complete item 9.
   Example:

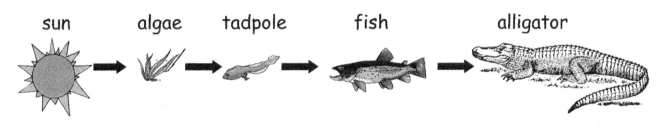

9. _____ uses energy from the sun to make food during photosynthesis. Algae is food for the _____. The tadpole is food for the _____, and the fish is food for the _____.

10. The transfer of energy from organism to organism begins with energy from the _____.

Lesson #8

The Flow of Energy from the Sun

All living things need **energy** to live and to grow. You already know that plants get their energy from the sun. Green plants are called **producers** because they are the only living organisms that can **produce** (make) their own food. Besides sunlight, plants also need air, water, and nutrients to survive and grow.

1. Plants are called _____ because they can make their own food.

Look at the drawings below.

2. Which drawings represent producers? _____

A   B   C   D   E

Animals **consume** (eat) plants, and that is how they get energy. Animals are called **consumers**. Another word for consume is *use* – animals *use* plants as a source of energy. All animals must consume plants or other animals, and plants get their energy from the sun. Therefore, all living things – plants and animals – depend upon sunlight to survive.

3. All animals are (producers / consumers) because they eat plants or other animals.

4. All living things depend on sunlight because _____.
    A) Earth rotates around the sun.
    B) Sunlight is more powerful than artificial light.
    C) Plants use sunlight to make food, and animals eat plants.
    D) The sun is at the center of our solar system.

Simple Solutions© Science   Level 5

5. The results of her experiment are inconclusive. What might a scientist do next?

     A) Conduct the experiment again.
     B) Change the information in the data chart.
     C) Make up a different conclusion.
     D) Re-write the hypothesis to match the outcome of the experiment.

6. Underline any terms that name **natural resources**.

     water   money   trees   soil   libraries

7. Underline any examples of precipitation.

     sleet   hail   clouds   snow   glacier   rain

8. Write the steps of the scientific method in order.

     A) _____
     B) _____
     C) _____
     D) _____
     E) _____
     F) _____

9. What is the purpose of recycling?

     A) to reduce pollution
     B) to save energy
     C) to conserve resources
     D) all of these

10. List three materials that can be recycled.

_____   _____   _____

Lesson #9

Variables and Constants

Before beginning any experiment, a scientist must define variables. A **variable** is any factor that can vary or change in an experiment. The **independent variable** is the one thing that the scientist can change. All other variables should remain **constant** (unchanging).

For example, if the scientist is testing the effect of a fertilizer on bean plants, the fertilizer would be the **independent variable**. The scientist may add fertilizer to one plant but not another. However, both plants should be the *same type of plant*. They should get the *same amount of water and sunlight*, and they should have the *same air around them*. The type of plant, amount of water, and location are **constants**.

Let's say one of the plants is in a room where people are constantly smoking cigarettes. This would add a new variable. If one plant begins to wither, the scientist would not know if the wilting is because of the fertilizer or the cigarette smoke.

1. A **variable** is a factor that _____.

    may change   never changes

2. Another word for **constant** is _____.

    independent   changeable   unchanging

3. What is the difference between a variable and a constant?
    A) There is no difference between the two.
    B) A constant does not change, but a variable can change.
    C) There are many variables in an experiment.
    D) An experiment always has a constant but may not have a variable.

4. Plants can make their own food. That is why we call plants _____.

    consumers   producers   herbivores   food chains

5. What two things are needed for every experiment?

    materials list   goggles   procedures   report   timer

Simple Solutions© Science　　　　　　　　　　　　　　　　　　　　　　　　　　　　Level 5

6. What makes all animals **consumers**?

   A) Animals need shelter to survive.
   B) Animals make their own food.
   C) Animals need more energy than plants.
   D) All animals eat plants or other animals.

7. Underline three things that plants need to make their own food through photosynthesis.

   oxygen　　　water　　　carbon dioxide　　　sunlight　　　shade

8. What is a landfill?

   A) a large hole filled with trash
   B) a place where plastics are recycled
   C) a large canyon carved out by moving water
   D) a type of garden

9. Which scientific instrument would you use to study the night sky?

   thermometer　　　microscope　　　hand lens　　　telescope

10. Animals with backbones are called _____.

    upright　　　invertebrates　　　anthropoids　　　vertebrates

Lesson #10

Ramón's Experiment

Ramón was told that rechargeable batteries last longer than disposable batteries. He used a battery-operated radio to test the running time of each type of battery. The rechargeable batteries were more expensive. But the disposable batteries lasted three times as long as the rechargeable batteries.

1. Which of these is the best explanation for Ramón's results?

    A) Ramón's experiment was a failure.
    B) The rechargeable batteries were not fully charged.
    C) Rechargeable batteries cost too much.
    D) Disposable batteries cost too little.

2. Ramón still wants to find out whether or not rechargeable batteries last longer than disposable batteries. What should he do next?

    A) Repeat the experiment with a new set of disposable batteries and a fully charged set of re-chargeable batteries.
    B) Write a conclusion that says disposable batteries last longer than re-chargeable batteries.
    C) Conduct an experiment with lithium and alkaline batteries.
    D) Repeat the experiment using a digital camera instead of a radio.

3. Earth's surface (below the soil) is made of solid rock called _____.

4. Which type of animal has hair or fur and feeds its young with milk from the mother?

    reptile   mammal   bird   amphibian   fish

5. Some animals have outer colors and patterns that hide or disguise themselves. What is this called?

    camouflage   hibernation   migration   reproduction

6. The scientific method begins with a _____.

The Food Chain: Producer, Consumer, Decomposer

You know that **producers** are plants that make their own food through photosynthesis. And **consumers** are animals that eat plants. When an animal dies, microorganisms called **decomposers** break down the animal's body and allow nutrients back into the soil. Plants use the nutrients to make food.

Decomposers are single-celled organisms like bacteria and protists. Other decomposers are fungi (mushrooms) and earthworms.

Decomposers put nutrients back into the soil.

Producers (plants) use nutrients from the soil to make food.

Decomposers break down the bodies of dead animals.

Consumers use the energy they get from producers to live and grow.

7. Bacteria, fungi, and earthworms are examples of _____.

8. A food chain always starts with a _____.
   producer   consumer   herbivore   omnivore

9. Our _____ is everything around us.
   season   environment   atmosphere   hydrosphere

10. A place where trash is buried is called a _____.
    dump   landfill   natural resource   mine

Lesson #11

Control Group

In an experiment, there may be a **control** along with the item that is being tested. In the experiment described in Lesson #9, the plant that does not get the fertilizer is the control. The other plant is treated with the variable (fertilizer). <u>The control in an experiment is the thing that does not receive the experimental treatment.</u>

1. The steps in a scientific procedure should be very specific. Which of these would be the best way to list a procedure for the experiment described above?
   A) Water the plants often.
   B) Place the plants in good locations.
   C) Keep the plants moist.
   D) Give each plant 20 ml of water every other day.

2. The item or group that does not receive the experimental treatment during an experiment is called the _____ group.

       hypothesis   control   variable   independent

Look at the drawings below.

3. Which drawings represent consumers? _____

4. Which represent producers? _____

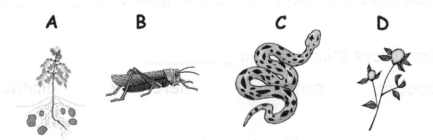

    A   B   C   D

5. Broken down rocks are carried away by wind, water, or moving ice in the process of _____.

       hibernation   glacier   erosion   migration

6. A decomposer is a consumer. Its job is to break down dead organic material to feed itself and to return nutrients to the soil. Which of the organisms illustrated below are decomposers?

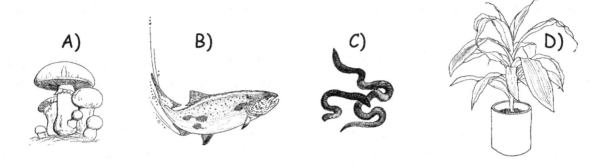

7. What does a decomposer consume?

   dead plant leaves   animal feces   dead animals   all of these

8. Which instrument is used to measure how hot or cold something is?

   barometer   ruler   thermometer

9. Which of the following causes erosion?
   A) moving water
   B) evaporation
   C) condensation
   D) humidity

10. According to laboratory safety rules, one thing you should never do in a science lab is _____.
    A) look into the lens of a microscope
    B) sniff or taste chemicals
    C) write in a notebook
    D) wear plastic goggles

Lesson #12

A Food Chain is Part of a Food Web

A **food web** is a series of food chains all linked together. Scientists use the term *food web* because all organisms are connected to many other organisms. A frog may eat a fly, but a fly will also eat a decomposing frog. A food web shows how energy is transferred from the sun and between organisms in an ecosystem.

Every simple food chain begins with **producers**; they use energy from the sun to make food during photosynthesis. Consumers come next. There are three levels of consumers: **Primary consumers**, also called **herbivores**, are plant-eaters. **Secondary consumers**, also called **carnivores**, eat other animals. **Tertiary consumers**, also called **third-level consumers**, eat animals that eat other animals.

Secondary and tertiary consumers may be carnivores or omnivores. Carnivores only eat meat (other animals); **omnivores** eat both plants and animals. Remember, primary means *first*; secondary means *second*, and tertiary means *third*. **Scavengers** are a special type of carnivore. Scavengers – like crows, vultures, and maggots – feed on the remains of dead animals. They help to clean up the environment by getting rid of decaying animal flesh. When scavengers are finished, decomposers take over.

Here is an illustration of a food web:

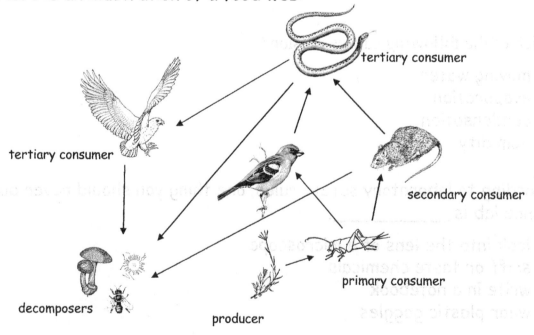

All of the arrows could point to the decomposers because they consume all of the other members of the food chain <u>but only after they have died</u>. Whatever decomposers don't use, they put back into the soil. Producers (plants) need the nutrients that decomposers put back into the soil.

Simple Solutions© Science   Level 5

1. Which animal gets its energy from the mouse?
        blue bird   hawk   cricket   mushroom

2. Which organism shown in the food web is an herbivore?
        hawk   snake   mouse   cricket   fly

3. Kevin enjoys fresh vegetables like broccoli, carrots, and cucumbers. He eats berries, apples, bananas, and just about any other kind of fruit. Kevin also eats hamburgers, grilled chicken, and fish sticks. Which type of consumer is Kevin?
        herbivore   carnivore   omnivore   none of these

Write each word next to the hint that describes it.

    omnivore   secondary consumer   tertiary consumer
    producer   primary consumer   decomposer

4. beginning of the food chain   _____

5. eats plants only   _____

6. eats animals that eat plants   _____

7. eats both plants and animals   _____

8. eats animals that eat other animals   _____

9. breaks down organic matter   _____

10. Why are some of the arrows in the food web pointing to the mushroom, bacteria, and fly?

_____

Simple Solutions© Science   Level 5

Lesson #13

What is a Cell?

All organisms (living things) are made of cells. A cell is called the basic unit of life because it is the smallest part of a living thing. Some whole organisms are just one single cell, but most organisms are made of a countless number of cells. Cells can only be seen through a microscope.

What are the Parts of a Cell?

Plant and animal cells are very similar. A **cell membrane** protects the cell. Both plant and animal cells take in nutrients through the cell membrane. This membrane also allows waste products to go out of cells. Plant and animal cells both have cytoplasm. **Cytoplasm** is like a jelly that fills the cell. It protects everything in the cell, keeping it healthy. Every plant and animal cell also has other smaller parts including a **nucleus** which controls the cell's activities, as well as **vacuoles** and **mitochondria**. Each part of a cell has a job to do, and all the parts work together to keep the organism alive.

1. Label this diagram of an animal cell. Find this diagram in the Help Pages.

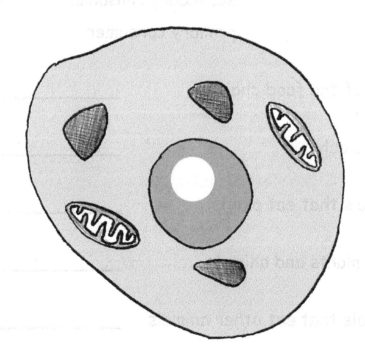

2. A _____ is the basic unit of life.

3. Producers use water, light, and carbon dioxide to make food

   in a process called _____.

4. Which word describes a primary consumer?

    omnivore   carnivore   producer   herbivore

5. Which two of these words may describe a secondary level consumer?

    herbivore   carnivore   producer   omnivore

6. Which two of these words may describe a tertiary or third-level consumer?

    herbivore   carnivore   producer   omnivore

7. When a tornado warning has been issued, what should you do?

    nothing   stand under a tree   go indoors   get out of the house

Dr. Hager's Experiment

Dr. Hager suspects that students who do their homework while watching TV don't perform as well as students who do homework in a quiet area. He assigns one group of students to complete their homework in a room with a TV that is tuned to a popular program. This is the experimental group. A second group of students is given the same homework assignment but is assigned to work in a quiet room. The students working in the quiet room are the control group.

8. One constant (stays the same) in this experiment is the _____.

    students   work area   assignment   television program

9. The independent variable (one thing that is changed) in Dr. Hager's experiment is the _____.

    work area   assignment   type of desk   writing instrument

10. Which of these results would verify (support) Dr. Hager's findings?

    A) Both groups of students perform equally well on the assignment.
    B) The control group performs well; the experimental group performs poorly.
    C) The control group performs poorly; the experimental group performs well.

Lesson #14

How do Plant and Animal Cells Differ?

Although they are very similar, plant and animal cells do have some differences. Outside of the cell membrane, a plant cell also has a **cell wall**. The cell wall helps all the plant cells stick together, and this supports the plant. Plant cells also have chloroplasts. Chloroplasts contain chlorophyll which makes plants green. **Chloroplasts** make food for the plant. Animal cells do not have chloroplasts.

1. Label this diagram of a plant cell. Find this diagram in the Help Pages.

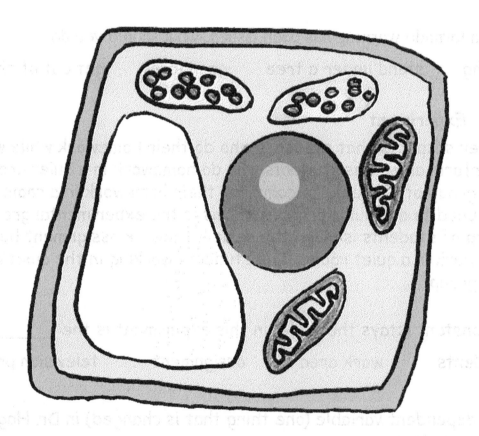

2. Another word for a living thing is an _____.

3. The smallest part of a living thing is a _____.

4. Consumers get their food from producers in a transfer of energy known as _____.

    the food chain   the water cycle   metamorphosis   deforestation

5. The word *carnivore* means _____.

    meat-eater   plant-eater   producer   animal

6. An *herbivore* will never eat _____.

    plants   animals   seeds   nuts

Repeated Trials

Scientists want to be sure that the results they get from an experiment are accurate. They want to **verify** (show) that the results of the experiment were not just a coincidence. The only way to be sure is to repeat or **replicate** the experiment. Sometimes an experiment is replicated several times – even hundreds of times – before a scientist or group of scientists is willing to say that the results are **conclusive** (definite). Scientists are looking for **evidence** – facts that support their conclusions. If there is not enough evidence, the results are **inconclusive** which means that nothing has been proven.

7. Why do scientists often repeat an experiment?

    A) They need to verify their results.
    B) Most scientists enjoy spending time in the lab.
    C) Scientists are usually paid by the hour.
    D) It is important to use up all materials in a science lab.

8. Which scientific instrument would you use to examine the surface of a rock?

    thermometer   barometer   magnifying glass   rock hammer

9 – 10. Match each word with its meaning.

   ___ verify   A) proving nothing

   ___ replicate   B) definite

   ___ conclusive   C) repeat

   ___ evidence   D) show or support

   ___ inconclusive   E) facts

Lesson #15

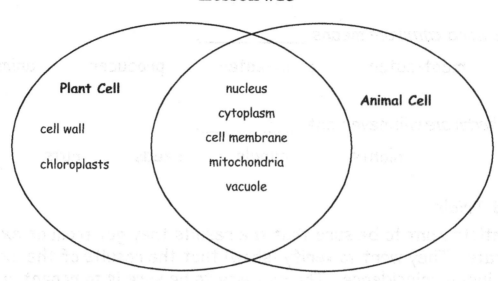

1. According to the Venn diagram, only _____ cells have a cell wall.

2. The diagram shows that plants and animals _____.
   A) have cells that are completely different
   B) have cells that are very similar
   C) are one-celled
   D) have chloroplasts in their cells

3. The word **distinction** means difference. What is one distinction between plant cells and animal cells?

   _____

4. _____ feed upon decaying organic matter (dead plants and animals).
   A) Scavengers and decomposers   C) Herbivores
   B) Primary consumers   D) Predators

5. One type of harmful pollutant is _____.

   hazardous waste   petroleum   soil   compound

6. Rock is broken down into soil through _____.

   weathering   migration   natural resources   conservation

Classification

When you **classify** (organize), you group things together according to their characteristics. For example, when you organize your school supplies, you may put all of your notebooks together in one place. You may keep your textbooks all together and your pens and pencils in a separate place. This makes everything easy to find when you need it.

Another way to organize is to keep all the things you need for one subject together. You may decide to keep your ruler, pencil, math text, and math notebook all in one place because these are all the things you need for math class. As you can see, there are many ways to classify.

7. To classify means to _____.

All of the items pictured below are tools. How can you group these tools? Give two broad titles.

8. _____

9. _____

10. One way that scientists classify animals is to put them into two groups: vertebrates and invertebrates. What major characteristic do vertebrates have? (Check the Help Pages if you're not sure.)

_____

Lesson #16

Classification

You may group things together according to their size, shape, or how they are used. Where do you keep your ruler? Maybe you keep your ruler inside a notebook since it is about the same length as a notebook. Maybe you keep your ruler with your pens and pencils since a ruler is a tool, and writing utensils are tools. Size, shape, and use are all examples of **characteristics**.

Scientists do not always agree about how to classify organisms. And scientists are making new discoveries all the time. That's why some scientists say there are five kingdoms and others say there are six. There are also slightly different names for the kingdoms. However, all scientists classify organisms according to their characteristics.

1. Scientists _____ organisms according to their characteristics.

2. Size, shape, and use are examples of _____.

3. How can you group these items? Give two or three broad titles.

   _____

4. Which tool would be the best instrument to use to measure the height of a plant's stem?

   thermometer   hand lens   ruler   beaker

5. Plants make food in a process called **photosynthesis**. What does a green plant use to make food during photosynthesis?

   A) soil, fertilizer, and water
   B) seeds, water, and energy from sunlight
   C) water, carbon dioxide, and energy from sunlight
   D) oxygen, energy from sunlight, and carbon dioxide

Simple Solutions© Science Level 5

More about Animal Cells

In animals, cells make up tissue and organs. **Tissue** is simply a group of cells. An **organ** is a group of tissues. A **body system** is a group of organs working together. For example, muscle cells make muscle tissue, and muscle tissue makes a heart which is an organ. The heart is part of a system. It works together with blood, arteries, veins, and tiny capillaries to make up the circulatory system.

> muscle cell ➡ muscle tissue ➡ heart ➡ circulatory system

Where do cells come from? They come from other cells through **cell division**. When the nucleus divides itself, one cell becomes two cells. Through cell division, all the parts of the original cell show up in the new cell.

Use these terms to complete the sentences below.

> cell   nucleus   tissue   organ   systems

6. Cells work together to form _____.

7. Every plant and animal cell has a _____ that directs all of the cell's activities.

8. Organs work together to form body _____.

9. A group of tissues working together forms a(n) _____.

10. A _____ is the smallest unit or part of a living thing.

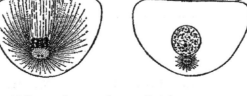

When the nucleus divides, one cell becomes two cells.

33

Simple Solutions© Science   Level 5

Lesson #17

Classification

There are millions of organisms on Earth, so scientists need a way to keep them all organized. Living things are grouped together according to their similarities. There are at least five **kingdoms** (some scientists identify six). The members of one kingdom are all animals. Another kingdom has only plants, and another has only single-celled, microscopic organisms. All living things belong to one of the Kingdoms:

Kingdom	Illustration	Description	Examples
Monera (Bacteria)		one-celled; no nucleus; may absorb or make own food	green sulfur bacteria, purple bacteria, acidophilus
Protista (Protists)		one-celled with nucleus; absorb or make own food; some protists are multi-cellular	amoeba, diatom, euglena, algae, paramecium, protozoa
Fungi		many-celled; absorb food from their environment; some fungi are unicellular	mushrooms, puffballs, mold, yeast, mildew, toadstools
Plants		many-celled; cells contain chloroplasts & can make food	trees, flowers, shrubs, grasses, cacti, seaweed, ferns, moss
Animals		able to move; many-celled; feed on plants and animals	monkeys, birds, fish, octopus, elephants, cats, spiders, humans

1. Which kingdoms have organisms that can make their own food?

    animal   plant   fungus   bacteria   protist

34

Simple Solutions© Science   Level 5

2. Which kingdoms have single-celled organisms?

    animals   plants   fungi   bacteria   protists

3. Which kingdoms have many-celled organisms?

    animals   plants   fungi   bacteria   protists

4. A cactus, a fern, and a tree belong to the same kingdom because ____.

    A) they all grow outdoors
    B) they all make their own food
    C) they all die in cold weather
    D) they all need water to survive

5. Which pair of organisms is most closely related?

    panda – house cat   pine tree – bacteria   algae – spider

6. What is one distinction (difference) between bacteria and protists?

    _____

7. The illustration shows that cells make more cells through _____.

    cell division   cell maintenance   ph

8. Which of these is not part of the scientific method?

    A) questioning   C) observation
    B) experiments   D) science fiction

9. When a liquid becomes a gas or vapor, the process is called _____.

    A) evaporation   C) precipitation
    B) melting   D) freezing

10. Which scientific instrument would you use to examine skin cells?

    forceps   spring scale   balance   microscope

35

Lesson #18

Ecosystems

All of the living and nonliving things interacting and affecting each other in a certain area make up an **ecosystem**. Ecosystems come in all sizes. Both a puddle and a rainforest can be ecosystems. There may be many different populations living in each of these ecosystems. A **population** is simply a certain group of the same kind of organism. For example, the puddle ecosystem may have a population of bacteria or tiny insects. The rainforest may have a population of tree frogs, ferns, or fungi. Earth supports many different ecosystems, and these ecosystems support many different types of organisms.

Organisms can only thrive in healthy ecosystems. A healthy ecosystem meets an animal's needs for **air, food, water,** and **shelter**; it meets a plant's needs for **sunlight, air,** and **water**. Ecosystems also have two other features: carrying capacity and waste disposal. **Carrying capacity** is the population size that an ecosystem can support without damaging the ecosystem. For example, only a certain number of frogs can live in a pond. What will happen if the population of frogs gets too large? There won't be enough food, and some of the frogs may die of starvation. The pond may become polluted because of all the waste produced by the frogs and the bodies of dead frogs. Oxygen in the water may be used up, and that would cause other organisms to die.

What about **waste disposal**? An ecosystem disposes of its own waste by constantly recycling organisms and non-living materials. If an ecosystem is not able to properly dispose of waste, it may become polluted. Pollution can choke the life out of an ecosystem. **Decomposers** (like fungi, bacteria, protists, and earthworms) are an ecosystem's recyclers. They break down organic matter and put nutrients back into the ecosystem.

1. Organisms and the environment they live in make up a(n) _____.

2. _____ recycle organic material by breaking it down and returning nutrients to the ecosystem.

3. The size of a population that can live in a certain ecosystem is called the system's _____.

4. A group of the same kind of organism is called a _____.

5. Which pair of organisms is most closely related? (If you're not sure, look back at the chart in Lesson #17.)

   A) turtle and seaweed
   B) yeast and mold
   C) monkey and tree
   D) elephant and purple bacteria

6. One distinction between a plant cell and an animal cell is _____.

   A) a plant cell has a cell wall; an animal cell does not
   B) an animal cell has a nucleus; a plant cell does not
   C) plant cells contain cytoplasm; animals cells do not
   D) animal cells have a cell membrane; plant cells do not

7. Which of these is a producer?

   groundhog   corn plant   soil   sun

8. A _____ is the basic unit of life.

   atom   seed   cell   consumer

9. Which eats only only plants?   herbivore   carnivore   omnivore

10. Which two of these words may describe a tertiary or third-level consumer?

    herbivore   carnivore   producer   omnivore

This decaying tree provides nutrients for the mushrooms growing on it. The mushrooms are decomposing the dead tree and will put nutrients from the tree back into the ecosystem.

Lesson #19

1. A scientific investigation usually <u>begins</u> with a _____.

    question   conclusion   experiment   result

2. Name the five kingdoms. (See the chart in Lesson #17.)

    _____   _____

    _____   _____

    _____

3. Scientists use _____ as a way of grouping or organizing things into easy-to-manage groups.

    investigation   hibernation   classification   observation

4. Look back at the **Five Kingdoms Chart**. Some of the protists and bacteria are like plants in one way. What is it?

    _____

5. A distinction is a difference. One **distinction** between plant and animal cells is _____.

    A) plant cells have chloroplasts and animal cells don't
    B) an animal cell has a nucleus
    C) plant cells contain cytoplasm
    D) animal cells have a cell membrane

6. An _____ is made up of all of the living and nonliving things interacting and affecting each other in a certain area.

    arena   ecosystem   omnivore   anemometer

Simple Solutions© Science   Level 5

Scientists call animals with backbones **vertebrates**. Vertebrates are classified into five major groups: mammals, fish, birds, amphibians, and reptiles. Use the chart to answer the questions that follow.

Vertebrates				
	Body Covering	Most Common Type of Reproduction	Breathing	Examples
Mammals	hair or fur	babies born alive; feed young with milk	lungs	horses, whales, cats, humans
Fish	scales	lay eggs in water	gills	salmon, trout, catfish, carp
Birds	feathers	lay eggs with hard shells	lungs	ostrich, cardinal, hummingbird
Amphibians	moist skin	lay eggs in water	gills, then lungs	frogs, toads, salamanders
Reptiles	dry, scaly skin	lay eggs	lungs	snakes, turtles, alligators

7. How do scientists classify a bat – is it a bird or a mammal?

   A) A bat is classified as a bird because a bat can fly.
   B) A bat is classified as mammal because its body is covered with fur, and it feeds its young with milk.
   C) A bat is neither a bird, nor a mammal.
   D) A bat is both bird and mammal.

8. What is one **distinction** between reptiles and fish?

   _____

9. Why are dolphins, porpoises, and whales classified as mammals, and not as fish?

   A) Dolphins, porpoises, and whales breathe with lungs.
   B) Dolphins, porpoises, and whales can live on land, and fish can't.
   C) Dolphins, porpoises, and whales are too big to be fish.

10. An alligator spends much of its time in the water. Why is an alligator classified as a reptile instead of a fish or an amphibian?

    A) Alligators are much larger than most amphibians.
    B) An alligator lives its whole life breathing though lungs, but an amphibian starts out with gills.
    C) Alligators are carnivores.

Lesson #20

Just the Facts, Please

Scientists use observation to gather information. To observe means to use your senses of sight, smell, taste, hearing, and touch. (Remember, science safety rules say **you should never eat or drink anything during a science activity**.) Scientists look for data like facts and statistics, but they avoid jumping to conclusions or making assumptions. An assumption is like an opinion – it cannot be proven.

1. Which are facts and which are assumptions? Write **F** for fact or **A** for assumption.

    A) _____ The tomato plants are healthy.

    B) _____ The tomato plants have dark green leaves.

    C) _____ The stem is 7 cm in length.

    D) _____ The hamster likes to run on the exercise wheel.

    E) _____ The hamster ran on the exercise wheel for 3 minutes.

2. Put these steps of the scientific method in order. Number from 1 – 6.

    _____ Gather data from your experiment.

    _____ Conduct an experiment.

    _____ Share conclusions based on the results of your experiment.

    _____ Formulate a hypothesis.

    _____ Start with a problem or question.

    _____ Do some research to gather more information.

3. What is **waste disposal** in an ecosystem?

    A) an ecosystem's ability to recycle dead plants and animals
    B) the process completed by scavengers and decomposers
    C) a characteristic of an ecosystem that allows it to support life
    D) all of these

4. List the two main functions of a plant root.

    _____   _____

Simple Solutions© Science   Level 5

5. Animals that only eat other animals are called _____.

   herbivores   carnivores   omnivores   producers

6. Animals that eat both plants and other animals are called _____.

   herbivores   carnivores   omnivores   producers

7. Animals that only eat plants are called _____.

   herbivores   carnivores   omnivores   producers

8. Match these.

   ___ primary consumers   A) make their own food

   ___ secondary consumers   B) eat plants only

   ___ tertiary consumers   C) eat animals that eat plants

   ___ producers   D) eat animals that eat other animals

Organisms interact with each other in their environment. You already know that organisms depend upon their ecosystems for air, food, water, and shelter. Plants get nutrients from the soil and light from the sun. Animals hunt for food or hide from predators. Decomposers munch away at decaying things and put nutrients back into the soil. These are all examples of how organisms interact with each other in their environment.

9. Which of these is an example of how organisms interact with each other in their environment?

   A) flooding   C) erosion
   B) the predator-prey relationship D) all of these

10. Consumers get their food from producers in a transfer of energy known as _____.

    A) the food chain   C) the water cycle
    B) metamorphosis   D) deforestation

Lesson #21

Complete the crossword puzzle using the bolded words you see in the text.

Maria Sibylla Merian, Naturalist and Botanic Illustrator (1647 – 1717)

Maria Sibylla Merian was a life-long observer and illustrator of nature. A fascination with plants, flowers, and bugs motivated Maria to become a great contributor to the field of **entomology** (the study of insects). As a young girl, Maria began to investigate and draw the animals and plants that she saw in her **environment**. She worked alongside her step-father who painted beautiful flowers. Maria was an excellent painter as well, but she was much more interested in the bees and beetles that took their **nourishment** from the plants that her father painted. Butterflies and moths were also very fascinating, especially when Maria discovered that these insects move through the phases of **metamorphosis**. She used only a hand lens to observe the little organisms since microscopes were just being developed. Still, Maria loved to observe the tiny **chrysalis** (cocoon), the fuzzy coat of the caterpillar, and the tiny antennae of each winged creature as it grew through its stages of development.

As an adult, Maria continued to study insects by collecting and raising them in containers in her kitchen. She classified insects according to their diets and their behavior and documented her work carefully. Because Maria was able to show each stage of butterfly metamorphosis in rich detail, her work was greatly admired by **botanists** and **zoologists**. During Maria's lifetime, it was very unusual for a woman to be a scientist or to travel on a long journey on her own. But Maria did both of these things and much more. She traveled from her home in Holland to Suriname, in South America because she wanted to observe live **specimens** in their natural **habitat**. Maria spent two years studying and drawing the unique creatures that she found on the island. In 1705 she published a very important book called <u>Metamorphosis of the Insects of Surinam</u>. Maria also published many other books, and she sold numerous paintings of animals, insects, plants, and flowers. Maria was a remarkable woman who became famous and widely respected by other scientists and artists.

These are examples of Maria's drawings.

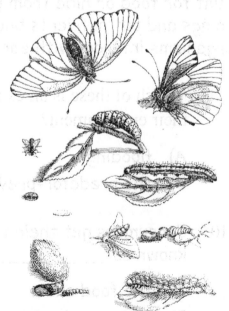

Simple Solutions© Science

Level 5

Across
1. plant scientists
6. everything around us
7. food and water
9. cocoon
10. animal scientists

Down
2. samples
3. the study of insects
4. country in South America
5. change of phases
8. the place where an organism lives

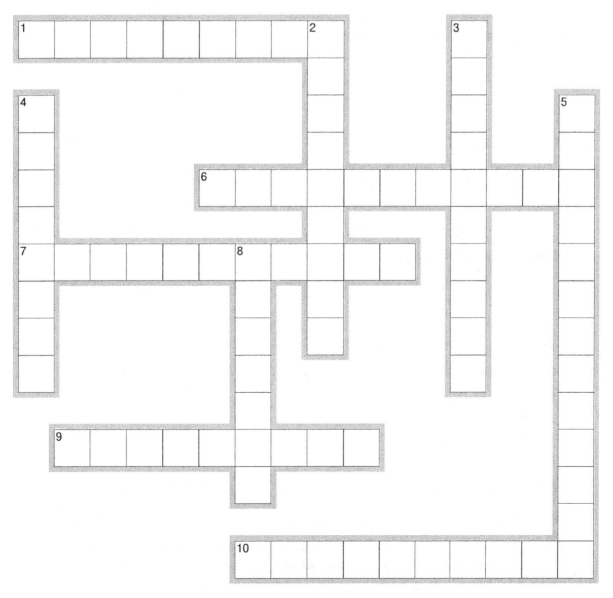

Lesson #22

1. The _____ is a step by step approach to finding answers to questions about the world around us.

2. Which of the following are characteristics of a <u>healthy</u> ecosystem?

   A) pollution and oxygen
   B) ample carrying capacity and a waste disposal system
   C) overpopulation and limited food supply
   D) nice weather and good location

3. The roots of some plants store food. Which group lists plants that store food in their roots?

   A) corn, lettuce, strawberries
   B) cacti, deciduous trees, tulips
   C) radishes, carrots, turnips

4. A _____ is the smallest unit or part of a living thing.

5. Which kingdoms have organisms that can make their own food?

   animals   plants   fungi   bacteria   protists

6. One example of how **organisms interact with one another** in an environment is a _____.

   food web   weather event   volcanic eruption

7. One example of how **organisms cause change** in an environment is ____.

   A) scavengers and decomposers working on road kill
   B) people building highways
   C) plant roots breaking up rock
   D) all of these

8. What is one distinction between predators and scavengers?

   A) Predators may eat animals and plants; scavengers only eat plants.
   B) Predators hunt live animals; scavengers seek out dead organisms.
   C) Predators eat only flesh; scavengers eat only plant material.
   D) Predators live in the wild; scavengers do not.

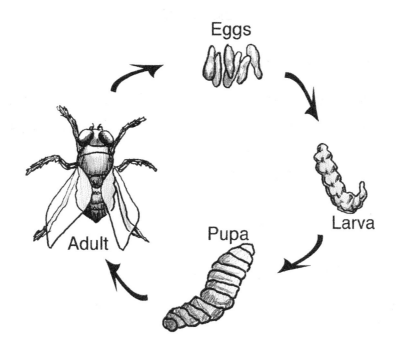

9. The diagram shows the stages of complete _____.

   photosynthesis   hibernation   migration   metamorphosis

10. Besides houseflies, which of these organisms go through the stages shown in the diagram?

    chicken   mosquito   alligator   butterfly   dolphin

Construction of highways, shopping malls, and other structures forces organisms out of their natural habitats. These organisms either die off or are forced to move to another area.

45

Lesson #23

1. What is the best way to test a hypothesis?

    A) Conduct an experiment.
    B) Create a data chart.
    C) Ask a scientist.
    D) Take a guess.

2. Producers use water, light, and carbon dioxide to make food in a process called _____.

3. Which kingdoms have single-celled organisms?

    animals   plants   fungi   bacteria   protists

4. Which of the following are adaptations that animals use for survival?

    hibernation   camouflage   migration   mimicry   all of these

How do Organisms Cause Change in their Environment?

The inside of a human mouth is warm, moist, and full of food residue. That makes the mouth an ideal **environment** for the many micro-organisms that live there. Some of these micro-organisms are harmless while others can be very harmful.

One particular type of bacteria – *streptococcus mutans* – lives in plaque on the surface of teeth. When a person eats sweets, the bacteria begin to feed on the sugar. As the bacteria breaks down sugar, they produce acids. These acids destroy the minerals in tooth enamel, and the tooth begins to decay after several of these "acid attacks." As the tooth decays, a cavity develops.

5. In the example described above, how does the organism called *streptococcus mutans* change its environment?

    _____

    _____

6. What do the bacteria need that the environment in a person's mouth provides?

    food   shelter   water   all of these

7. Describe something that would decrease or limit the amount of *streptococcus mutans* that could live in your mouth.

   _____

   Humans are organisms that cause change in their environment. Humans destroy the habitats of many living things by clearing land for housing, shopping centers, and roads. Deer, raccoons, rabbits, squirrels, insects, plants, and even micro-organisms either die or are forced into new habitats. Some will adapt and survive; others will not. Humans also change the environment by releasing toxic waste from factories and cars into the air. The chemicals in these fumes can cause acid rain which burns plants and contaminates the water in rivers and lakes.

8. What is one way that humans cause change in their environment?

   _____

9. Two acres of forested land is cleared in order to build a large shopping center. What might happen to the organisms that lived in the forest?

   A) Some will die.
   B) Some will adapt and continue to live in the shopping center area.
   C) Some will leave and find a new habitat.
   D) All of the above are possible.

10. What is a population?

    A) the decaying matter in an ecosystem
    B) a certain group of the same type of organism
    C) many different kinds of animals living together in a place
    D) a field of various kinds of plants

Air pollution makes it difficult to breathe, and it can cause contamination of water in rivers and lakes.

Lesson #24

Abiotic Factors Cause Change in Environments

Changes in environment are nothing new; the Earth has been changing for many millions of years. Over time, many species of plants and animals have become endangered or gone extinct because of changes in their environments. Sometimes humans are the cause of these changes, and sometimes the changes are caused by shifts in climate, earthquakes, volcanic eruptions, and other catastrophic events.

Environmental change has a chain-like or "domino effect." One change – even a small one – can lead to many other changes. That is because each organism interacts with so many other organisms and abiotic factors in its environment. **Abiotic** means non-living. Some abiotic factors are soil, sunlight, air, water, and climate. If an environment is flooded, the water supply may become polluted, and all living organisms in that environment will be affected. Plants and soil may be washed away. Animals that depend on the plant life may starve, and other animals that prey on those animals would go hungry as well.

1. Which of these is an abiotic (nonliving) factor that may cause change in an environment?

    flood   bacteria   invasive species   overpopulation

2. A(n) _____ starts its life in the water and lives on land as an adult.

    reptile   mammal   amphibian   avian

3. A _____ is the basic unit of life.   atom   seed   cell   consumer

4. Which of the following may decrease an ecosystem's **carrying capacity**? Choose all that apply.

    severe flooding   rotation of Earth   air pollution

5. Consider the organisms pictured in the food web. Explain three changes that may occur as a result of a long-term drought in their environment.

    _____

    _____

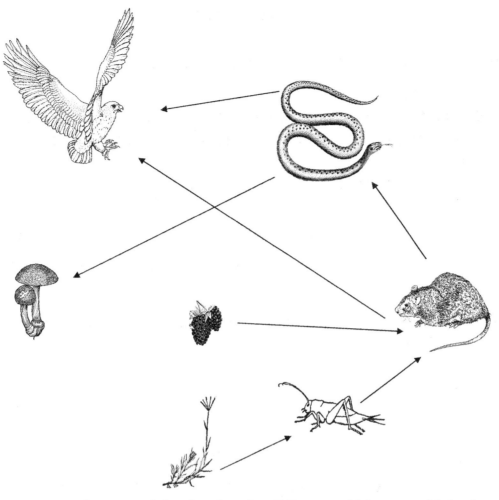

6. The hawk is at the top of the food web. What could happen if the hawk population became too large?

    A) Most of the mice and snakes may be wiped out.
    B) The population of crickets may increase.
    C) The grasses that feed the crickets may be depleted.
    D) All of these could happen if there were too many hawks.

7. The mouse pictured in the food web is a(n) _____.

    carnivore   herbivore   omnivore   producer

8. The mushrooms pictured in the food web represent _____.

    producers   decomposers   predators   prey

9. Which organism pictured is a primary consumer? _____

10. Which organism is a secondary consumer? _____

Lesson #25

1. Plant stems have two main functions. What are they?
   A) gather sunlight and make food
   B) support the plant and carry water and nutrients to leaves
   C) make food for the plant and reach water supplies deep within the soil
   D) produce carbon dioxide and make seeds

2. Plant cells and animal cells contain a _____ which directs cell activity.

       mitochondrion   seed   cell   nucleus

3. According to the Venn diagram below, what are three similarities between reptiles and amphibians?

   _____

4. Which group goes through the stages of metamorphosis?

       reptiles   amphibians   both   neither

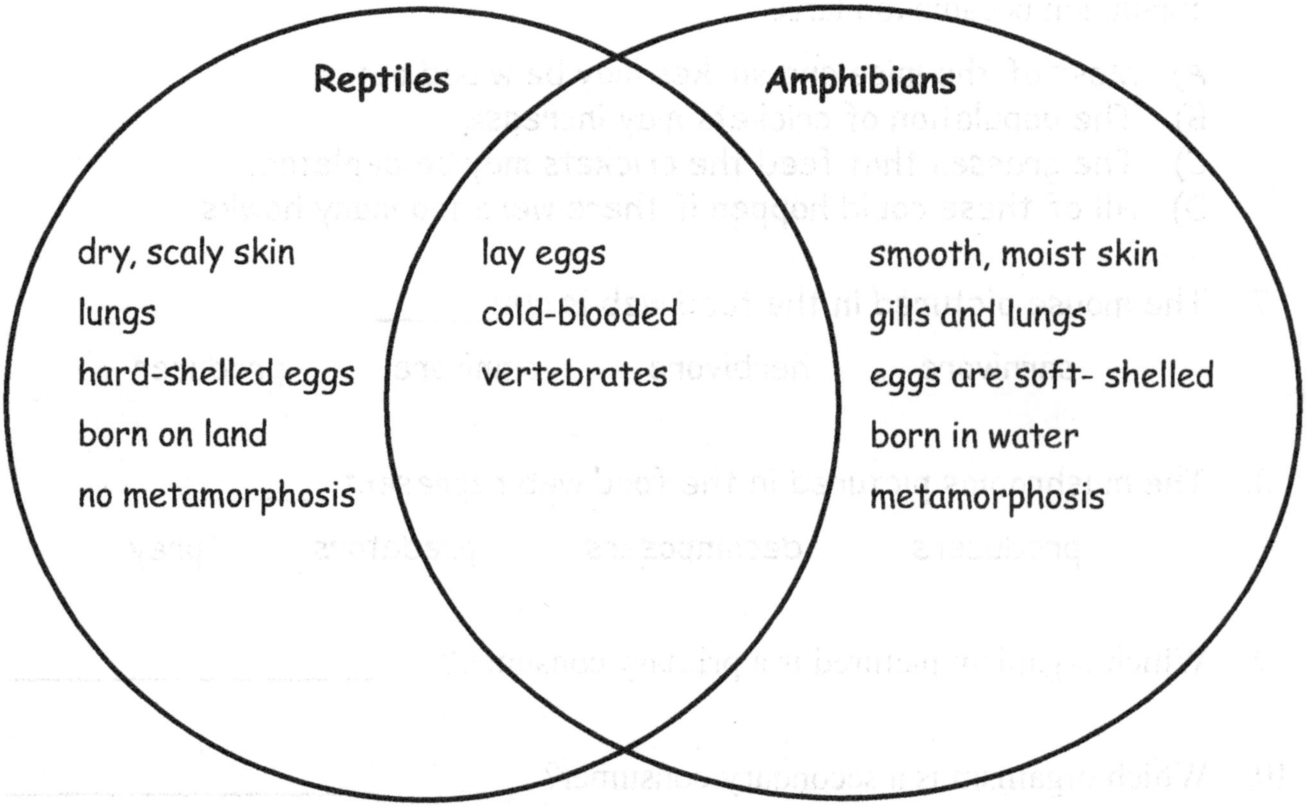

Reptiles
- dry, scaly skin
- lungs
- hard-shelled eggs
- born on land
- no metamorphosis

Both
- lay eggs
- cold-blooded
- vertebrates

Amphibians
- smooth, moist skin
- gills and lungs
- eggs are soft-shelled
- born in water
- metamorphosis

5. A cactus, a fern, and a tree *belong to the same kingdom* because _____.

   A) they all grow outdoors
   B) they all make their own food
   C) they all die in cold weather
   D) they all need water to survive

6. How might an earthquake cause change in an ecosystem?

   A) Some animals' homes might be destroyed.
   B) Some animals may be driven from their habitat.
   C) Plants and animals may be killed.
   D) All of the above

7. What effect do weathering and erosion have on the environment?

   A) no effect
   B) rock is broken down
   C) thunderstorms increase
   D) temperatures decrease

8. Most erosion is caused by _____.

   the sun   wind   moving water   insects

9. A huge, slow-moving mass of ice that causes erosion by scraping against the surface of the Earth is a _____.

   bulldozer   boulder   mineral   glacier

10. _____ are materials carried away by moving water.

    Sediments   Snow and ice   Ecosystems   None of these

Weathering and erosion create interesting rock formations.

Simple Solutions© Science   Level 5

Lesson #26

A **food web** is a series of food chains all linked together. Study the food web illustration. Then answer the questions that follow.

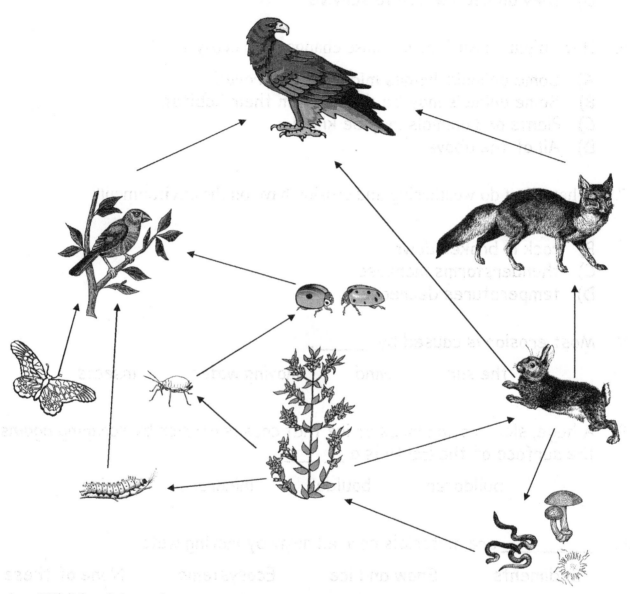

1. Which of these is a **secondary consumer**?

    worms   fox   rabbit   all of these

2. Which of these is a **tertiary consumer**?

    fox   ladybug   caterpillar   hawk

3. The flowering plant represents _____.

    decomposers   producers   consumers   herbivores

4. The caterpillar, butterfly, aphid, and rabbit are all _____.

   omnivores   primary consumers   decomposers   carnivores

5. The small bird, hawk, fox, and ladybugs are all _____.

   primary consumers   carnivores   decomposers   herbivores

6. Which of these is demonstrated by the food web?
   A) Each organism is connected to many other organisms.
   B) Organisms interact with each other in an ecosystem.
   C) All life depends upon a transfer of energy from the sun.
   D) All of these

7. What would likely happen if the population of foxes in this ecosystem decreased significantly?
   A) The hawks would become extinct.
   B) The population of rabbits would increase.
   C) The population of ladybugs would decrease.
   D) There would be no decomposers.

8. The food web shows an arrow going from the rabbit to the decomposers. Which of the following is true?
   A) All plants and animals are broken down by decomposers once they die.
   B) Only animals are broken down by decomposers.
   C) Only plants are broken down by decomposers.
   D) Only dead rabbits are broken down by decomposers.

9. Which pair of organisms is most closely related?

   frog/fly   bacteria/grass   mushroom/ant   bird/moss

10. What is **carrying capacity**?
    A) the ability to properly recycle waste in an ecosystem
    B) the population size that can live in an ecosystem without causing damage to that ecosystem
    C) a variety of different populations living together in an ecosystem.

Lesson #27

The Changing Earth

Earth is constantly being shaped. In fact, Earth's surface has been changing for many billions of years. Some of the **processes** that shape the Earth are weathering, erosion, and the shifting of plates. **Weathering** is the wearing away of rock by water, wind, and ice. **Erosion** is the movement of sediment (broken down rock) by water, wind, and ice. The upper part of the mantle is soft, and the **tectonic plates** (also called continental plates or oceanic plates) float over this layer. The plates are like pieces of a giant puzzle. They fit together closely, but they also move a little. The movement of these plates is so slight that it can not even be noticed by humans. However, this slight movement causes bumping and pulling which create landforms that we can see on the surface of Earth. Shifting plates cause earthquakes and volcanic eruptions. Over many billions of years, all this shifting and flowing have formed mountain ranges, islands, and Earth's other surface formations.

Sections of Earth's crust, called plates, fit together like pieces of a puzzle. The movement of plates shapes the Earth's surface.

1. Most of the processes that shape the Earth take millions of years. Which of these can cause sudden, extreme changes?

    A) weathering and erosion
    B) earthquakes and volcanoes
    C) glaciers and icebergs
    D) moving water, wind, and ice

2. Water, wind, and ice cause _____.

    erosion   weathering   changes in Earth's surface   all of these

3. Little pieces of broken down rock are called _____.

    sediment   lava   plates   erosion

Simple Solutions© Science   Level 5

4. Which of the following is most likely to have created the mountains pictured here?

   A) wind erosion
   B) action of plant roots
   C) movement of tectonic plates

5. What does inconclusive mean?

   incorrect   absolute   uncertain   failed

6. When a liquid becomes a solid, the process is called _____.

   evaporation   precipitation   melting   freezing

7. Making food for the plant is the main function of the _____.

   leaves   stem   roots   sap

8. Rain, snow, sleet, and hail are all examples of _____.

   A) evaporation   C) freezing
   B) precipitation   D) sublimation

9. What do plants cells have that animal cells do not have?

   A) mitochondria   C) cell wall
   B) cell membrane   D) cytoplasm

10. Which scientific instrument would you use to measure the length of a leaf?

    A) spring scale   C) tape measure
    B) thermometer   D) balance

    Although the Earth's crust is its thinnest layer, the crust contains many layers of soil, and subsoil containing organic matter, sediments, sand, minerals, and rock. Beneath all of these layers is bedrock.

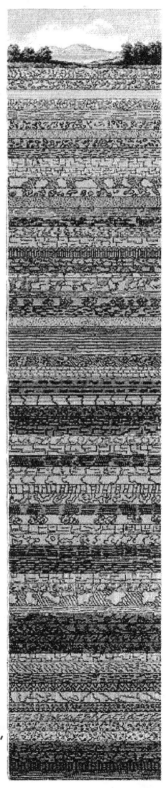

55

Lesson #28

Use the diagram to answer the first three questions.

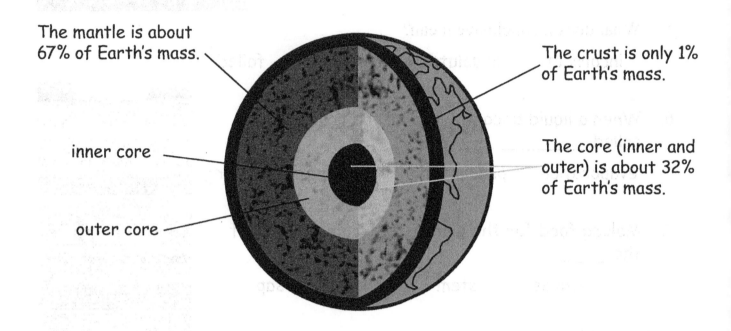

1. Earth is made up of layers. The crust (outer layer) is the _____ layer of the Earth.

    thickest   thinnest   deepest   biggest

2. The innermost layer of the Earth is called the _____.

    core   mantle   crust   continents

3. The Earth's _____ is much thicker than the crust or the core.

4. Sections of Earth's crust that fit together like puzzle pieces and move very slowly are called _____.

    mantle   core   plates   continents

5. Organisms and the environment they live in make up an _____.

    area   atmosphere   ecosystem

Simple Solutions© Science　　　　　　　　　　　　　　　　　　　　　　　　Level 5

6. Which of these can cause earthquakes and volcanoes?

   weathering　　　erosion　　　plate movement　　　severe weather

7. Which pair of organisms is most closely related? (See Lesson #17)

   centipede/moss　　　protozoa/ fish　　　puppy/worm　　　moss/whale

8. Cells make more cells by _____.

   dividing　　　synthesizing　　　hibernating　　　migrating

9. Match each word with its meaning.

   ___ beginning of the food chain　　　A. omnivore

   ___ eats plants only　　　B. primary consumer

   ___ eats animals that eat plants　　　C. secondary consumer

   ___ eats both plants and animals　　　D. tertiary consumer

   ___ eats animals that eat other animals　　　E. producer

   ___ breaks down organic matter　　　F. decomposer

10. If decomposers are the "last link" in the food chain, what is the first link?

    carnivores　　　herbivores　　　plants　　　humans

Decomposers are the last link in the food chain because they break down dead organic material and put nutrients back into the soil for plants.

Lesson #29

1. Read each clue. Write **M** if it describes a mammal or **R** if it describes a reptile. Write **B** if the clue describes both a mammal and a reptile.

    A) _____ vertebrate

    B) _____ skin covered with hair or fur

    C) _____ lays eggs

    D) _____ breathes with lungs

    E) _____ warm-blooded

2. About how long does it take for the moon to go through its phases?

    21 days   365 days   30 days   7 days   24 hours

3. Water becomes a solid through a process called _____.

    evaporation   freezing   precipitation   sublimation

4. Underline words that name examples of natural resources.

    plastic   tree   newspaper   air   cars   water

5. All cells contain a _____.   nucleus   cell wall   chloroplast

6. Which one of these allows a plant cell to make food during photosynthesis?

    chloroplasts   nucleus   cell wall   flowers   fruit

How Old is that Fossil?

Scientists study fossils to learn more about what Earth was like millions of years ago. The Earth's crust has many layers, and the deepest layers are the oldest. Just like the layers of a cake, the part that came first is at the bottom. After that, each layer is a little newer or younger. And the layer on the top goes on last, so it is the newest. When scientists uncover a fossil, they try to analyze the age of the rock layer where the fossil was buried. This is one way of determining the age of the fossil.

 Sometimes rock layers are disturbed by earthquakes. Molten rock pushes its way up through cracks in the Earth's crust, and then the layers are not in their original order.

Shifted layers

7. The imprint or remains of things that lived long ago are called _____.

   animals   reptiles   fossils   artifacts

8. Which of these is not a fossil?

   A) dinosaur footprint   C) petrified wood
   B) insect in amber   D) leaf sample

9. A plant or animal that no longer exists anywhere is considered to be _____.

   endangered   rare   extinct   isolated

10. The oldest fossils will most likely be found in the _____ layers of Earth's crust.

    deepest   middle   most shallow   most moist

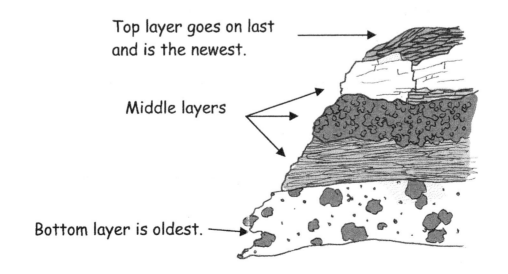

59

Lesson #30

Rocks and Minerals

If you like to collect rocks, you know that there are many different kinds, and there are hundreds of different minerals. A **mineral** – unlike a rock – has a specific chemical make-up. A mineral is a naturally occurring, **inorganic** (never living) solid. Rocks are made of minerals, but <u>minerals are not rocks</u>. Some examples of minerals are diamonds, coal, salt, and gold. Minerals have certain properties or characteristics that help scientists to identify and classify them. Color is one of the properties of a mineral. Hardness is another. Some minerals glow, others are magnetic.

1. How many different minerals are there?

    about a dozen   only a few   hundreds   four

2. Which of these is not a fossil?

    diamond   petrified wood   dinosaur footprint   trilobite imprint

3. Name two properties or characteristics of minerals.

    _____   _____

Complete the following sentences with words from the word bank.

- A) Evaporation
- B) Weathering
- C) Photosynthesis
- D) Minerals
- E) Plates

4. _____ is the wearing away of rock by water, wind, and ice.

5. _____ are naturally occurring inorganic solids.

6. _____ are sections of Earth's crust and mantle.

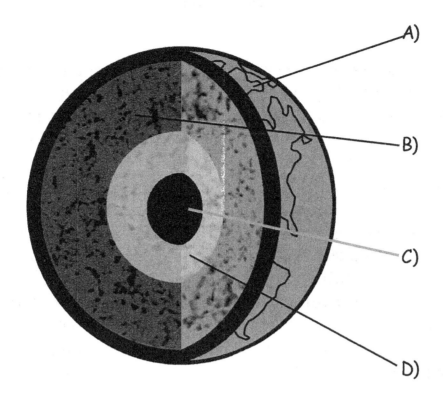

7. Label the diagram of Earth's layers.

   A)_____   C)_____

   B)_____   D)_____

8. Which layer is the thinnest?

   mantle   outer core   crust   inner core

9. What is the relationship between predator and prey?
   A) A predator hunts prey for food.
   B) Prey hunts predators for food.
   C) A predator lives inside its prey.
   D) Prey breaks down and eliminates the bodies of dead predators.

10. Grouping living things based on similar characteristics is called _____.

    species   hibernation   classification   kingdom

Lesson #31

The Rock Cycle

There are three types of rock: igneous, sedimentary, and metamorphic. **Igneous** rock is formed when molten rock cools and hardens. **Sedimentary** rock is formed when rock particles called sediments are bonded together by pressure over time. **Metamorphic** rock forms when high heat and pressure change the make-up of a rock into a new type of rock.

The three types of rock – igneous, sedimentary, and metamorphic – are constantly changing from one form to another in a process called the **rock cycle**. Igneous rock comes from magma that cools. The word *igneous* means "from fire." Weathering and erosion will cause igneous rock to break into bits and pieces. Under extreme pressure, these bits and pieces, called *sediments*, are cemented together to form sedimentary rock. High heat and pressure will change any kind of rock into metamorphic rock. The word *metamorphic* means "changing form." A metamorphic rock has changed into something completely different.

Choose terms from the word bank to complete the sentences below.

- A) igneous
- B) metamorphic
- C) weathering
- D) magma
- E) sediments

1. When rocks break apart, the bits and pieces are called _____.

2. Melted rock cools and hardens to form _____ rock.

3. With enough heat and pressure, any rock can be changed to a _____ rock.

4. Melted rock is called _____.

5. _____ is the process that breaks rock down into bits and pieces.

6. Which of these carry water and nutrients to the plant?

   roots   stem   leaves   flowers   fruit

Simple Solutions© Science   Level 5

The Rock Cycle

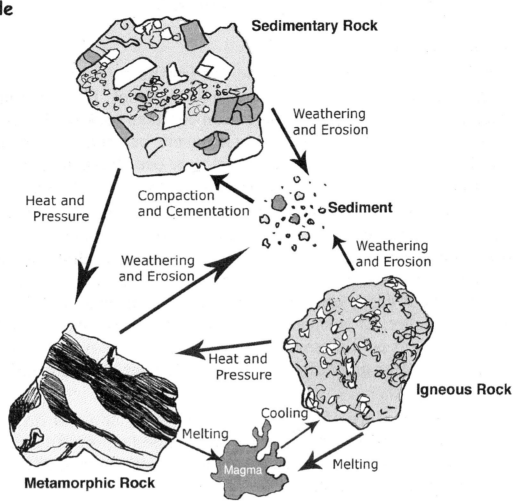

7. What are the four basic needs of animals?

   _____   _____

   _____   _____

8. Which kingdoms have many-celled organisms?

   animals   plants   fungi   bacteria   protists

9. Which of these are naturally occurring, inorganic solids?

   decomposers   minerals   plants   oceans

10. The imprint or remains of things that lived long ago are called _____.

Lesson #32

The Perfect Mix

Soil, a thin layer of material on the Earth's surface, is made up of rock particles, air, water, and decayed organic material. (Decayed plant and animal matter is also called **humus**.) Since soil contains rock particles, it contains minerals. (Remember, rocks contain minerals, but minerals are not rocks.) Plants have their roots in the soil, and they use these minerals, as well as other nutrients and water, for making food and for growing.

Soil is formed over a long period of time by the interaction of air, water, organisms, rocks, and chemicals. There are many different colors and **textures** (how fine or coarse the soil is) of soil. **Sand** has very large particles; it is loose and feels coarse or rough. Sand helps soil to drain well, but it dries out easily and does not hold many nutrients. **Silt** is a soil with smaller particles. It is smooth and powdery and is made of sediments. **Clay** has the smallest particles and is heavy, especially when it is wet. Clay can hold a lot of water and nutrients, but it does not allow air and water to move freely through it.

Loam is a mixture of soil that contains sand, silt, and clay along with humus, water, and air. As you know, plants need air, water, and nutrients to live and grow. Loam is a soil that can hold all of these, and that is why it is best for growing plants.

1. Why is loam the best type of soil for growing plants?

   _____

2. Why would it be difficult to grow crops in sandy soil?

   A) Sand does not hold water and nutrients very well.
   B) Sand is too heavy.
   C) Sand is hard to find.
   D) None of these

3. Frogs, toads and salamanders are all examples of which?

   reptiles   birds   amphibians   mammals

4. Which scientific instrument would you use to examine skin cells?

   balance   forceps   spring scale   microscope

5. How do plants use soil?

    A) Plants anchor themselves by putting down roots into the soil.
    B) Plants get nutrients from the soil.
    C) Plants get water from the soil.
    D) All of the above

6. Match each term with its clue.

    ___ metamorphic   A) from fire

    ___ igneous   B) from bits and pieces

    ___ sedimentary   C) changing form

7. Mark shuffles across the floor wearing socks on his feet. When he touches the light switch, he gets a little shock. What type of electricity gave him the shock?

    lightning   static   magnetic   current

8. If the "like" poles of two magnets are put next to each other, _____.

    the poles will attract   the poles will repel   the poles will do nothing

9. Which scientific instrument would you use to study the night sky?

    telescope   microscope   hand lens   forceps

10. Use information from the paragraphs to identify each soil type in the chart below.

Soil Texture		
Soil Type	Particle Size	Description
A)	smallest	heavy; does not allow flow of water and nutrients
B)	small	smooth and powdery; made of sediments
C)	very large	loose and rough; provides good drainage
D)	medium	mixture of all other soil types; rich farming soil

Simple Solutions© Science　　　　　　　　　　　　　　　　　Level 5

Lesson #33

1. Cells work together to form _____.

   　　parts　　　tissue　　　nuclei　　　mitochondria

2. What does *abiotic* mean?

   　　organic　　　living　　　natural　　　non-living

3. Why would it be difficult to grow plants in clay?

   A) Clay gets extremely hot.
   B) Clay does not hold enough water for plants.
   C) Clay does not have any nutrients.
   D) Clay does not allow air and water to move through it.

Igneous rock is formed in two places: underground and above ground. When **magma** (melted rock) gets trapped in underground pockets, it cools, forming igneous rock. When a volcano erupts the magma flows out; then it is called **lava**. Cooled lava also becomes igneous rock. The movement of lava pushing its way to the surface is one of the processes that shape the surface of the Earth.

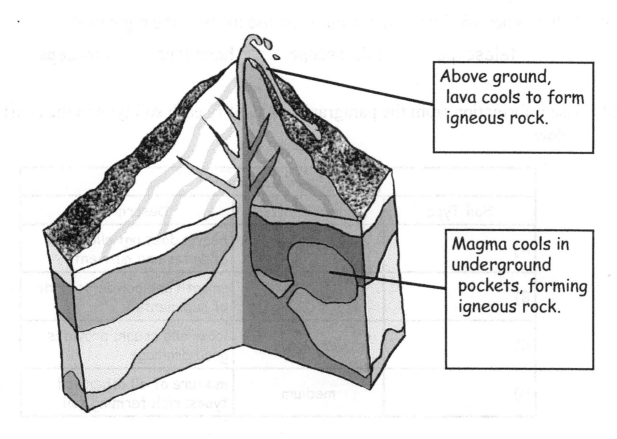

Above ground, lava cools to form igneous rock.

Magma cools in underground pockets, forming igneous rock.

4. Why is igneous rock sometimes called "fire rock?"

   _____

5. Soil is made up of many things. List three.

   _____   _____   _____

6. How do the organisms pictured interact with each other in their environment?
   A) Earthworms may provide food for a bird.
   B) Earthworms may decompose a bird's body when it dies.
   C) Both A and B
   D) None of the above

7. Name the first step of the scientific method.

   _____

8. Which of these can cause weathering?

   car exhaust   moving water   sunshine   photography

9. _____ are born in water, live near water, and lay their eggs in water.

   Birds   Mammals   Reptiles   Amphibians

10. Look at the illustration. In which layer would you expect to find the oldest fossils?

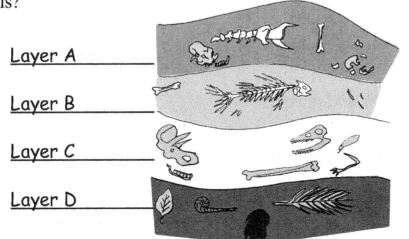

Lesson #34

1. The population size that can live in an ecosystem without causing damage to that ecosystem is called the system's _____.

    food chain   waste disposal   shelter   carrying capacity

2. Birds, fish, and humans all belong to the same **kingdom** because _____.

    A) they feed off of plants and animals   C) they are multi-cellular
    B) they do not produce their own food   D) All of the above

3. The illustration below shows that cells **reproduce** (make more cells). Cells do this by _____.

    adding   dividing   subtracting   multiplying

4. Write each word next to the hint that describes it.

    omnivore   secondary consumer   tertiary consumer
    decomposer   primary consumer   producer

    A) eats animals that eat other animals   _____

    B) breaks down organic matter   _____

    C) eats animals that eat plants   _____

    D) eats both plants and animals   _____

    E) beginning of the food chain   _____

    F) eats plants only   _____

Sedimentary rock forms over many thousands of years, as **sediments** (little pieces of broken down rock) are carried away by wind and water. These sediments sink to the bottom of lakes, rivers, and oceans. They settle down in layers; this settling is called **deposition** because the sediments are deposited onto the lowest surface. The sediments turn to rock as a result of the weight of these layers pressing down on one another.

5. What causes rock to break down into **sediments**?

    evaporation   cementation   condensation   weathering

6. Look at the animal pictured below. To which vertebrate group does the animal belong?

    reptile   mammal   amphibian   bird

7. Bits and pieces of rock that are carried by wind or water settle to the ground; this is called _____.

    weathering   deposition   fracturing   overflow

8. When a liquid changes to a gas, the process is called _____.

    condensation   evaporation   sublimation   melting

9. What are the four types of soil?

    _____   _____

    _____   _____

10. Choose the terms that name **natural processes** that shape the Earth's surface.

    weathering
    movement of tectonic plates
    erosion
    coal mining

Lesson #35

Use the diagram to answer the first three questions.

Arctic Food Web

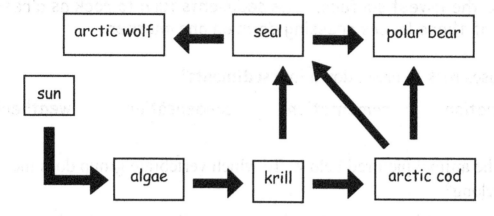

1. The seal gets its energy by eating _____ and _____.

2. If the population of krill suddenly decreased, what would likely happen?

   A) The population of arctic cod would decrease.
   B) The population of algae would decrease.
   C) The population of seal would increase.
   D) None of these

3. According to the food web diagram, which of the following are **tertiary consumers**?

   A) algae, krill, and cod   C) krill, cod, and polar bear
   B) wolf, seal, and polar bear   D) algae only

4. What three things do green plants need in order to make their own food?

   _____, _____, _____

5. How are rock particles formed?

           weathering   erosion   both of these

Metamorphic rock started out as igneous or sedimentary rock. Intense heat and enormous pressure transformed the rock into something new. Rock that is beneath many layers of Earth has many tons of weight pressing down on it. This causes a build-up of heat, and the combination of heat and pressure causes the rock to flatten and change into a completely new kind of rock. Remember, **meta** means *change*, and **morph** means *shape*.

6. What does *metamorphic* mean?

    solid   changing shape   completely new   heated and cooled

7. What causes an igneous or sedimentary rock to change into a metamorphic rock? Choose all that apply.

    heat   pressure   water   decaying plants

8. Bits and pieces of rock settle to the ground during the process of ____.

    evaporation   condensation   deposition   precipitation

9. What is humus?

    a mineral   inorganic material   decaying plants and animals

10. If you are performing an experiment, when is the best time to gather your data?

    A) before the experiment
    B) during the experiment
    C) after the experiment is over
    D) anytime

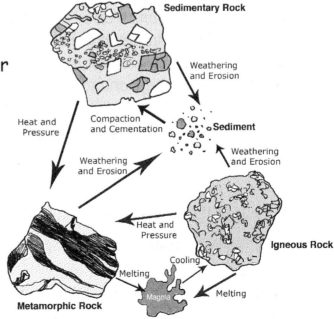

Lesson #36

Mindy's Experiment

Mindy bought a bunch of green bananas and separated them into three groups. She put one group in a brown paper bag, a second group on the kitchen counter, and the third group on a sunny windowsill. After 24 hours, Mindy compared all three sets of bananas to see how they had ripened.

1. Which of these hypotheses was Mindy probably testing?

    A) Bananas ripen faster in cool temperatures.
    B) Bananas ripen faster in warm temperatures.
    C) Fruit flies appear around very ripe bananas.
    D) Bananas ripen faster in a sunny area.

Mindy recorded her observations, and her data chart looked like this:

Banana Observation Chart			
Storage Area	After 1 Day	After 2 Days	After 3 Days
Paper Bag	slightly yellow		
Kitchen Counter	mostly green		
Windowsill	mostly green		

2. What should Mindy do next?

    A) Put the bananas back in the three locations for another day.
    B) Put the bananas in three new locations.
    C) Throw away the ripest bananas; keep the others where they are.
    D) Put all of the bananas in paper bags.

3. A _____ is the basic unit of life.

    atom   seed   cell   consumer

4. Which of these is a good conductor of electricity?

    rubber   dry wood   steel   Styrofoam

5. What does *metamorphic* mean?

    different   big mountain   changing shape   none of these

The Hydrosphere

Over 70% of Earth's surface is covered with water. Water is in oceans, lakes, rivers, streams, ponds, and puddles. Water is also frozen in glaciers and icecaps. There is water underground, and there is water vapor in the atmosphere. All of this water is part of the **hydrosphere**. **Hydro** means *water*, and **sphere** means *globe*, so the hydrosphere is all the waters of the Earth.

6. The hydrosphere is _____.

7. List three different places where the Earth's water is found.

   _____, _____, _____

8. The illustrated food chain shows that the spider is _____.

   A) a consumer because it eats flies
   B) a producer because it is a food source for the frog
   C) neither a consumer nor a producer
   D) both a consumer and a producer

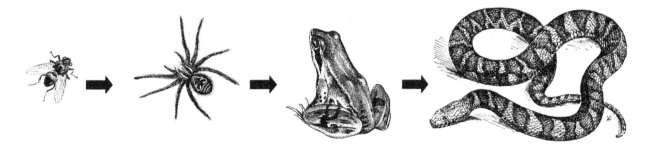

9. The illustration shows that the snake gets its energy from _____.

   A) frogs only
   B) flies only
   C) frogs and spiders only
   D) flies, spiders, and frogs

10. Which of these are forms of <u>solid</u> precipitation?

    glaciers   hail   rain   snow   water vapor

Lesson #37

Mineral Characteristics

Remember, a mineral is *a substance that occurs naturally;* that means it is found in the environment. A mineral is **inorganic** which means it is not living, and it never was alive. A mineral is a **crystalline** (kris tal ēn) **solid**. The structure of a crystal has a definite, repeating pattern.

All minerals have physical properties (characteristics) that help scientists to identify them. Some of these properties are color, streak, transparency, luster, and hardness. Minerals come in many **colors** including pink, green, and brown; some minerals are clear. **Streak** is the color of the mark that a mineral leaves when it is slid over a streak plate. The streak color is always the same for each particular mineral. **Transparency** means that light can pass through. Some minerals allow a lot of light to pass through them while others are **opaque** (allowing no light to pass through). **Luster** is the way light reflects off of the mineral. Gold and silver have a metallic luster. Other minerals are dull or glassy in luster. A mineral's **hardness** is determined by a **scratch test**. A harder mineral will scratch a softer mineral. Diamond is the hardest mineral, and talc is the softest. Talc is so soft that you could scratch it with your fingernail.

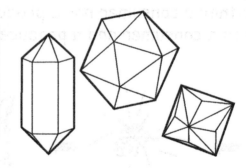

The shape of a crystal is smooth and symmetrical because its atoms are arranged that way.

1. List five of the physical properties of minerals.

   _____

2. What does inorganic mean?   hard   living   non-living   transparent

3. What is the hardest mineral? _____

4. What is the softest mineral? _____

5. What are some of the colors of minerals? _____

6. Most bears eat berries, plant roots, and seeds. They also eat insects and small animals such as mice. What kind of consumer is a bear?

   herbivore   carnivore   omnivore   producer

7. Which scientific instrument would you use to examine the surface of a rock?

   rock hammer   barometer   magnifying glass   thermometer

8. A first-level consumer eats _____.

   plants   other animals   consumers   both plants and animals

9–10. Look at the photographs below. Describe what will happen to the organisms living in the forest ecosystem as a result of the cutting of so many trees.

_____

_____

_____

Deforestation is the permanent destruction of forests and is caused by cutting down too many trees too quickly. Trees are cut to get wood which is used for fuel, building material, and paper. Forests are also cleared to make room for buildings and other uses.

Lesson #38

Experiment, Demonstration, or Model?

Your teacher may show a scientific concept or principle by performing a **demonstration**. And you may have even done a demonstration yourself – they can be really cool! Here are some examples of demonstrations:

- *showing* how Styrofoam peanuts dissolve in acetone
- *showing* how an object balances at its center of gravity
- *showing* how polystyrene will shrink when heat is applied

A demonstration is not the same as an experiment. Remember, an experiment is a test set up to answer a specific question. When you begin an experiment, you do not know what the result will be. The examples above **demonstrate** or *show* something that is already known.

A **model** is like a demonstration because a model is meant to show something.

- A dinosaur model helps us to see how the animal's body armor protected it from predators.
- A model of the digestive system shows how food travels through the body.
- A model of the solar system shows how planets orbit the sun.

Sometimes a model and a demonstration go together. For example, you can use a model of a volcano to show the volcano's structure *and* to demonstrate how a volcano erupts. You can demonstrate how to make slime with common household ingredients. The slime becomes a model if you use it to show off some of the properties of a polymer.

Which is it? Read each science project; write **E** if it is an experiment; write **D** if it is a demonstration, or write **M** if it is a model.

1. _____ Troy used clay to form a replica of the human heart. He labeled each part and explained how blood flows through the heart's four chambers.

2. _____ Pam floated a thin layer of oil over water in a saucer to show that the oil will not dissolve in water.

3. _____ Carla tested three brands of paper towel to see which absorbs the most liquid.

4. _____ Mark froze water into various shapes, and then let them melt to see whether the shape of the ice has any effect on the time it takes to melt.

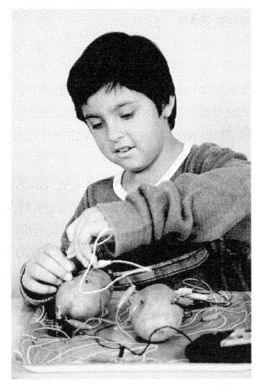

Student building a model

5. Which organism would not contain cells that have a cell wall?

    apple tree   fern   raccoon   cactus

6. Little pieces of broken down rock are called _____.

    sediments   lava   plates   erosion

7. A(n) _____ is an inorganic solid that has certain characteristics like color, streak, transparency, luster, and hardness.

    crust   mantle   igneous rock   mineral

8. An herbivore will never eat _____.

    plants   animals   seeds   nuts

9. Which organism is most likely to be at the bottom of a food chain?

    algae   crocodile   rabbit   grasshopper   skunk

10. Scientists group or classify living things based on similar _____.

    characteristics   kingdoms   species   migrations

Lesson #39

The Water Cycle

Every organism needs water, and water is continually moving above and below the surface of Earth. Water recycles itself in a process called the water cycle. The diagram shows that the water cycle has no beginning or end. It works like this:

Water on the Earth's surface evaporates, becoming **water vapor**. The atmosphere is filled with water vapor that circulates through the Earth's atmosphere. As water vapor cools, it condenses, forming clouds. When enough water condenses, heavy rain clouds give way to **precipitation** that falls to Earth's surface. (Remember, precipitation includes rain, snow, sleet, hail, fog, dew, or water in any form that falls to Earth's surface.) Water from precipitation or melting ice and snow sinks into the ground or runs into oceans, lakes, rivers, and other bodies of water.

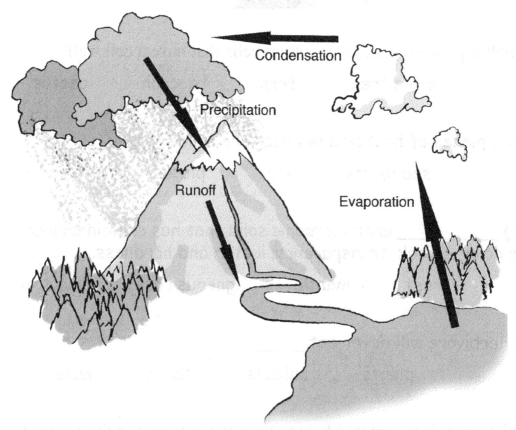

1. Water falls to the Earth in the form of rain, snow, sleet, and hail. This is called _____.

    precipitation   run-off   condensation   evaporation

2. Water turns to vapor in the process of _____.

    precipitation   run-off   condensation   evaporation

Simple Solutions® Science   Level 5

3. Where does the water cycle begin?

   A) precipitation   C) evaporation
   B) condensation   D) The water cycle has no beginning.

4. All of the waters of the Earth are known as the _____.

   atmosphere   lithosphere   hydrosphere   environment

5. List five of the physical properties of minerals. (See Lesson #37.)

   _____

   _____

6. What is humus?

   decaying plants and animals   mineral elements   inorganic material

7. Where do igneous rocks come from?

   sediments   cooled magma   mining   decayed plants

8. What are the four basic needs of animals?

   A) air, food, water, and shelter
   B) reproduction, safety, air, and food
   C) carbon dioxide, water, sunlight, and shelter
   D) ecosystems, food web, water cycle, and atmosphere

9. Mrs. Chen put a solution of yeast and sugar in a bottle and covered the bottle top with a balloon. Students observed the balloon expanding. Mrs. Chen wanted to show that the yeast solution produces a gas. Mrs. Chen was _____.

   A) doing an experiment   C) performing a demonstration
   B) building a model   D) gathering research

10. Which of these is one main function of a plant root?

    A) to absorb water and nutrients   C) to gather sunlight
    B) to produce seeds   D) to make food

Lesson #40

What Causes Condensation?

Condensation occurs when water vapor turns to liquid. When you see water droplets on the inside of a window or on the outside of a bottle of cold milk, you are observing this process. Condensation occurs when water vapor comes into contact with cooler air or a cooler surface. That is why a glass of ice-cold lemonade will form water droplets on its outside.

What Causes Evaporation?

A transfer of heat causes **evaporation**. When the sun heats up the air and the Earth's surface, water turns to vapor and begins to rise. If you observe a pot of boiling water, you will see steam (water vapor) rising from the boiling pot. If you leave a pan of water near a heater, the water will evaporate as the warmer air absorbs the water molecules.

What Causes Precipitation?

Water vapor condenses and forms clouds. **Precipitation** occurs when the condensed droplets become very large and heavy; they begin to fall to Earth. So, precipitation is really condensation falling to the Earth's surface.

1. Match these parts of the water cycle.

    ___ condensation   A) water falling to Earth's surface

    ___ precipitation   B) water vapor forming droplets

    ___ evaporation   C) water turning to vapor

2. About how long does it take for the moon to go through its phases?

    21 days   365 days   30 days   7 days   24 hours

3. What does *inorganic* mean?   living   non-living

4. When rocks break apart, the pieces are called _____.

5. Which would you use to measure the temperature outdoors?

    spring scale   tape measure   thermometer   balance

6. Which of these animals are invertebrates?

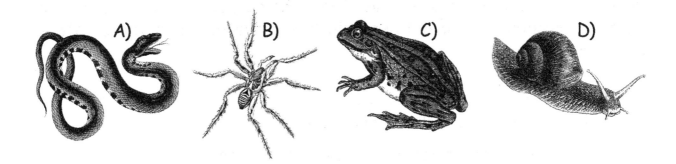

7. Many _____ begin their lives with gills; they develop lungs later on.

   invertebrates   mammals   birds   amphibians

8. What causes an igneous or sedimentary rock to change into a metamorphic rock? Check all that apply.

   _____ heat

   _____ pressure

   _____ water

   _____ decaying plants

Look at the diagram of Earth's layers.

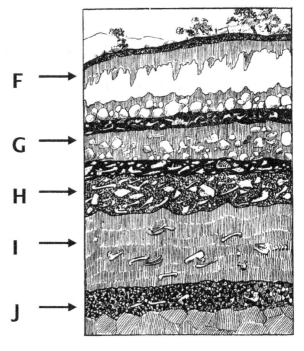

9. Which layer would contain the oldest fossils?

   _____

10. Which layer would contain the most recent fossils?

    _____

Lesson #41

For her science project, Diana used equal-sized samples of three different brands of toilet paper. For each sample, she soaked the paper in one liter of water for one hour. Then, she poured the paper and water mixture through a strainer to see which brand would leave the greatest residue.

1. Diana is trying to determine which brand of toilet paper is the most biodegradable. What should Diana do next?

    A) Add the toilet paper residue to a compost bin.
    B) Measure and record the mass of each residue sample.
    C) Put the paper residue into boiling water.
    D) Put the paper residue into the freezer.

2. Which of these best describes Diana's project?

    experiment   model   demonstration   report

3. What is the job of a cell nucleus?

    A) control the cell's activities   C) fill the space inside the cell
    B) hold the cell in place   D) none of these

4. In an animal's body, a group of tissues working together forms an ____.

    opening   organ   omnivore   octagon

5. A species that has died out is _____.

    endangered   extinct   hazardous   hibernating

6. Which of these belongs to the kingdom called protists? (Check the chart in Lesson #17 if you are not sure.)

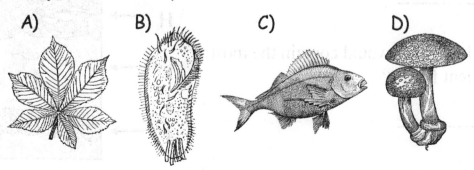

A)   B)   C)   D)

Surface water is water that is above ground. You can see it in lakes, rivers, streams, and oceans. But there is also water underground. Melting ice or snow and precipitation drain off the land; this is called **runoff**. Runoff soaks into the ground becoming **groundwater**, or it flows toward a body of water like a lake or ocean.

If too much water runs off too quickly, the ground may become saturated. That means no more water can soak in, and there will be flooding. If there are not enough plants to hold the soil in place, runoff causes erosion by carrying loose soil and sediments with it as it flows downhill.

7. Surface water is in _____, _____, _____, and _____.

8. Where does runoff come from?

   rain   snow   melting icecaps   all of these

9. How can runoff change and shape the Earth's surface?

   _____

10. Label the diagram. Use these words: **precipitation, evaporation, runoff, condensation**.

    A)_____

    B)_____

    C)_____

    D)_____

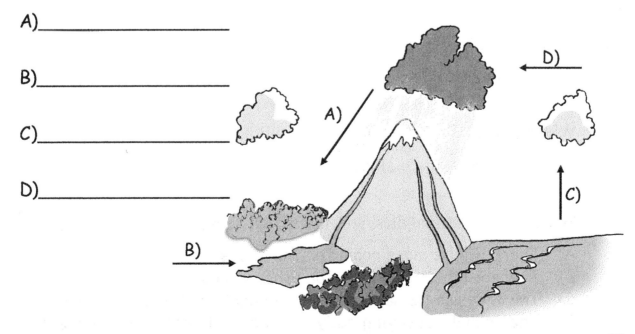

Lesson #42

Oceans are a very important part of the hydrosphere and the water cycle. Most of Earth's water is held in its vast oceans, and these oceans cover a large part of the surface of the Earth. Many tons of water are continually evaporating from oceans.

Precipitation falls on land and on bodies of water. Melting snow and runoff flow toward rivers that carry water to the oceans.

Some precipitation ends up in streams that flow to rivers, and the rivers flow back to the ocean.

Some of the precipitation that falls to Earth's surface soaks into the ground becoming groundwater.

Most groundwater flows back to the oceans.

1. What is the hydrosphere? (See Lesson #36.)

   _____

2. Most of the water on Earth is in _____.

   precipitation   oceans   groundwater   runoff

3. Which of these is not a type of precipitation?

   rain   sleet   rivers   snow   hail

4. When a forest animal dies, how might that help the trees in the forest?

   A) There will be more room for the trees.
   B) There will be more room for other animals.
   C) The tree will be healthier if no animal uses it for shelter.
   D) The animal's decaying body provides nutrients for the trees.

5. Which of these begins its life in water and later develops lungs and lives on land?

   reptile   bird   amphibian   mammal

6. A(n) _____ is the basic unit of life.

   atom   seed   cell   consumer

7. What do plant cells have that allow them to make their own food?

   stems   chloroplasts   membranes   leaves

8. Barry filled a glass with ice cubes and left it on the kitchen table. Several minutes later, Barry noticed that there were water droplets on the outside of the glass. What is the most likely explanation for this?

   A) Water was leaking from the glass.
   B) The glass was sweating.
   C) Water vapor was condensing on the cold surface of the glass.
   D) Precipitation was falling and adhering to the surface of the glass.

9. Fungi are not plants because fungi _____.

   A) cannot digest food.
   B) live on the ground.
   C) cannot be eaten.
   D) cannot make food

10. Which process would cause water droplets to accumulate on the leaf of this plant?

    A) photosynthesis
    B) vaporization
    C) condensation
    D) cell division

Lesson #43

1. What could happen to an ecosystem if all its plant life disappeared?
   A) Oxygen would be depleted.
   B) Animals would die of starvation.
   C) Both A and B could happen.
   D) None of the above

2. True or False?

   _____ Cells are microscopic.

   _____ Some living things are single-celled.

   _____ You can see the cells in a plant leaf by just looking at it.

   _____ Different kinds of cells do different jobs.

3. Some of the protists and bacteria are like plants in one way. What is it?

   _____

4. Which of these is an abiotic factor of an environment?

   mudslide overpopulation decomposers predators

Simple Machines

All machines are meant to make work easier for people. Most machines, like cars, power tools, computers, and dishwashers have many moving parts that work together. Simple machines are "simple" because they have only a few or no moving parts. Simple machines need a single force such as a push, a pull, or a lift to make them work. There are six simple machines: pulley, lever, wedge, wheel-and-axle, inclined plane, and screw.

5 – 10. Use information from the Help Pages to fill in the name of each simple machine described in the chart on the next page.

Simple Machine	Use	Example
5.	used to pry and to lift up heavy loads	
6.	makes it easier and faster to turn and to move things	
7.	allows things to be fastened together	
8.	used to raise and lower things	
9.	allows an object to slide; also used for cutting	
10.	makes it easier to move objects higher or lower	

Lesson #44

Jonas Salk, M.D., Medical Scientist & Developer of the Polio Vaccine (1914 - 1995)

During the first half of the twentieth century, babies often died from a horrifying illness called polio. Older children and adults also got the disease, and many recovered. But they were left with disabilities that made it difficult or impossible to walk. Parents were terrified that their children would catch the virus. No one seemed to know how to stop it. Even a U.S. President – Franklin D. Roosevelt – had been struck with the illness which left him partially paralyzed.

Meanwhile, a young Jewish girl named Dora left her home in Russia and moved to the United States. Sometime later, Dora married Daniel Salk, a young man whose parents had also emigrated from Northern Europe. The young couple lived in New York where they raised their three sons. The Salks were very poor, and neither Dora nor Daniel had much of an education. But they encouraged their three sons to study and work hard. The oldest boy, Jonas, was very bright, and he enjoyed spending time alone, just thinking. When he grew up, Jonas decided to become a lawyer. But Dora didn't think this was a good idea. She pointed out to Jonas that he could not even win an argument against his own mother! Jonas turned to medicine as his second choice.

Jonas Salk was the first member of his family to go to college. While he was attending medical school, Jonas began to research flu viruses, and he helped to develop successful **immunizations** (medicines that prevent disease) against the flu. In 1947, Jonas took a position at the University of Pittsburgh Medical School. There he began to work on a vaccine against the dreaded polio virus. By 1955, Jonas Salk had developed a vaccine that would prevent polio. People everywhere were relieved and grateful. After the development of the vaccine, every child could be protected against polio. Jonas Salk was a hero, and for the rest of his life, he continued to research ways to stop the spread of other diseases.

1. What causes polio?

    A) bacteria
    B) a virus
    C) exposure to the sun
    D) poverty

2. In the selection above, underline the meaning of **immunizations**.

Simple Solutions® Science Level 5

3. Which scientific instrument can be used to measure temperature?

   balance   barometer   beaker   thermometer

4. When the weather is cold, the inside of a car's windows may become foggy. What causes this?

   condensation   evaporation   leakage   precipitation

5. Which is the thickest part of the Earth?

   crust   mantle   outer core   inner core

6. Which is the thinnest layer of Earth?

   inner core   outer core   mantle   crust

7. Put a check next to two things you can do to prevent illnesses caused by a virus or bacteria.

   A) _____ wash your hands well and frequently

   B) _____ go to school or work everyday even if you have a fever

   C) _____ stay indoors away from fresh air and sunshine

   D) _____ strengthen your immune system by getting plenty of sleep

8. What is the relationship between plants and decomposers?

   A) Plants consume decomposers to get energy.
   B) Decomposers consume live plants.
   C) Plants use the nutrients that decomposers put into the soil.
   D) Decomposers eliminate the predators of many plants.

9. Which characteristic has to do with the way light reflects off of a mineral?

   transparency   streak   hardness   luster

10. What is a population?

   _____

Lesson #45

1. List five of the physical properties of minerals.

   _____

   _____

2. Which scientific instrument would you use to measure mass?

   dropper   balance   spring scale   microscope

3. What created this sea arch?

   A) a sculptor
   B) an earthquake
   C) moving water
   D) a tornado

4. This carnival ride is a good example of which simple machine?

   A) lever
   B) wedge
   C) wheel-and-axle
   D) inclined plane
   E) screw
   F) pulley

5. List two differences between plant and animal cells.

   _____

   _____

6. _____ is the wise use of natural resources.

7. Sponges, jellyfish, starfish, and worms are examples of _____.

         vertebrates   invertebrates

8. The word carnivore means _____.

     meat-eater   plant-eater   producer   animal

9. A plant's roots do which of the following?

   A) anchor the plant into the soil
   B) absorb nutrients and water from the soil
   C) both A and B
   D) none of the above

10. What is the best kind of soil for farming?

    A) silt
    B) sand
    C) clay
    D) loam
    E) a mix of sand and clay

Soil is a mixture. It contains bits of decaying plants and animals, broken down bits of rock, water, air, and minerals.

Lesson #46

1. Naomi needs to measure the water temperature in a fish tank. Which tool should she use?

    microscope   thermometer   balance   scale

2. Who or what is responsible for waste disposal in all ecosystems?
    A) the sanitation department
    B) scavengers, bacteria, fungi, worms
    C) producers and consumers
    D) the recycling department

3. Ecosystems come in all sizes. Which of these can be an ecosystem?

    rainforest   pond   desert   meadow   all of these

4. The innermost part of the Earth is its _____.

    mantle   crust   core   atmosphere

5. Explain what photosynthesis is and why it is important to all life on Earth. Use some of the words from the word bank.

animals	atmosphere	carbon dioxide	energy
food chain	nutrients	organisms	oxygen
process	producers	sun	water

    _____

    _____

    _____

    _____

6. If a magnet is put near a plastic bottle, the magnet will _____.

   A) attach to the bottle
   B) repel the bottle
   C) neither attract nor repel

7. A _____ is a very large mass of slow moving ice.

       fossil   sediment   glacier   river

8. Melted or liquid rock is called _____.

       magma   igneous   metamorphic   sedimentary

9. Which of these is another word for an inclined plane?

       pulley   screw   ramp   lever

10. A _____ is a simple machine that has grooved wheels and ropes. This machine makes it easier to raise and lower things.

A pulley is a rope, cable, or belt over a grooved wheel. It allows you to lift things up by pulling down on one end of the rope. This is how a flag is raised and lowered on a flagpole.

Lesson #47

Use the diagram to identify processes of the rock cycle. Complete each sentence with one of the phrases listed below.

A) Weathering and erosion   C) Compaction and cementation
B) Heat and pressure   D) Melting and cooling

1. _____ turn loose sediment into sedimentary rock.

2. _____ create igneous rock from magma.

3. _____ create metamorphic rock from other rocks.

4. _____ break up rock into bits and pieces and carry it away to be deposited in other places.

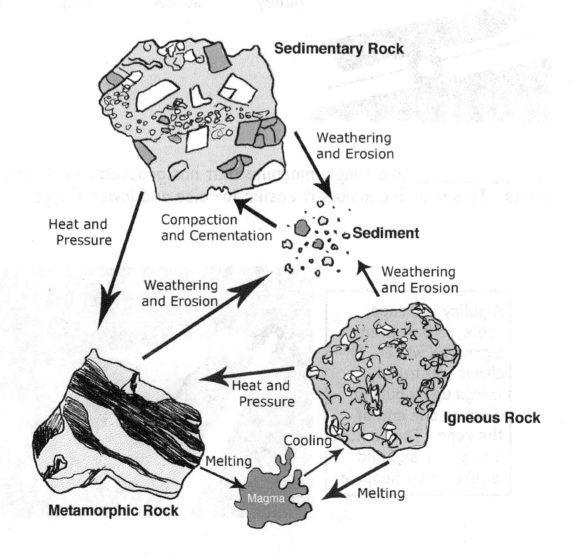

Simple Solutions© Science   Level 5

5. A knife blade is a good example of which simple machine?

      wheel-and-axle   pulley   ramp   wedge

6. Which two organisms belong to the same kingdom?

      blue jay   grass   protozoa   dandelion

7. What is the best way to test a hypothesis?

   A) conduct an experiment
   B) ask a scientist or teacher
   C) create a data chart
   D) guess

Name the Category

Complete the table below. Decide what the terms in each column have in common. Then write a title for each list, and fill in the terms that are missing.

8.	9.	10.
color		question
	wedge	
streak	wheel and axle	hypothesis
transparency	screw	
	inclined plane	data
magnetism		conclusion

Lesson #48

Darcy is planning a project for her school science fair. Read Darcy's description and answer the questions below.

- What I want to know: Which brand of cereal stays crunchiest?
- What I will do:

1) Soak 20 grams of cereal in 150 milliliters of water for 3 minutes.
2) Pour the cereal and water though a strainer into a measuring cup.
3) Measure the water that drains off and record the results in a data chart.
4) Repeat steps 1 – 3 with two more kinds of cereal.

1. Which of these is missing from Darcy's description?

    Question   Hypothesis   Procedure

2. What is the independent *variable* (the thing that changes) in Darcy's trials?

    amount of cereal   brand of cereal   amount of water   time

3. What is Darcy testing?

    A) how long it takes for cereal to dissolve
    B) how much liquid each brand will soak up in a given amount of time
    C) which cereal will be the tastiest
    D) how much water is needed to soak a given amount of cereal

Cereal Soaking Results

Brand	Amount of Leftover Water
Oatie O's	145 ml
Tasty Flakes	139 ml
Graham Crunch	30 ml

4. According to the results, which brand of cereal probably stays crunchiest?

    _____

5. Darcy wants to verify her results. What should she do next?

    A) Repeat the experiment with three other brands of cereal.
    B) Ask a friend to replicate the experiment and compare the results.
    C) Read the information on the labels of each box of cereal.
    D) Ask 50 people which cereal they think is the crunchiest.

6. Organs work together to form body _____.

        systems   cells   parts   hairs

7. Which pair of organisms is most closely related? (See Lesson #17.)

   A) tiger and centipede
   B) algae and tadpole
   C) snake and protozoan
   D) seaweed and starfish

8. Which of the following is an abiotic factor that may cause change in an environment?

    flooding   bacteria   plant roots   decomposers

9 – 10. Explain how energy from the sun is transferred from one organism to the next in the food chain pictured at the right.

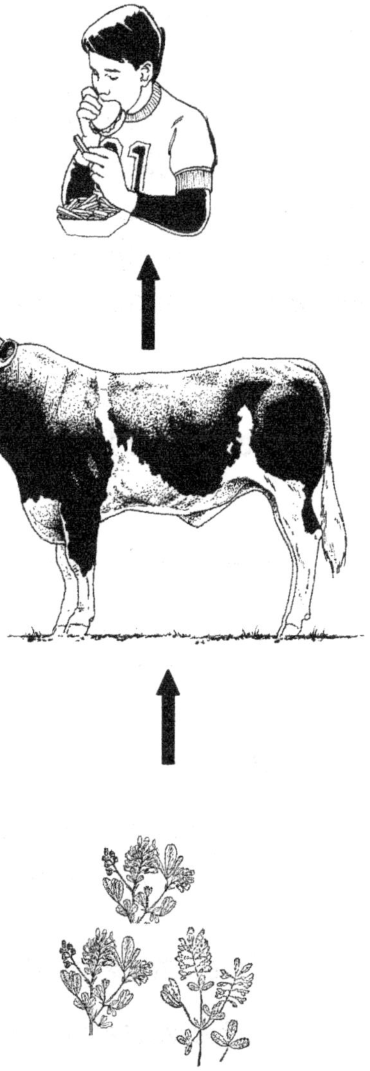

_____

_____

_____

_____

_____

_____

_____

Lesson #49

Planet Earth

Earth, as a planet, has everything that plants, animals, and other organisms need. What are these things? Well, first of all, our planet has a great location. Earth's **distance from the sun** allows it to be bathed in sunlight and warmth but not over-heated. Any organisms we know of could never survive in the extreme temperatures of some of the other planets. For example, Venus can have a surface temperature as high as 400° F, and Pluto (the dwarf planet) can have a surface temperature as cold as minus 400° F!

Earth also has an **atmosphere** that is one of its kind – as far as we know. Unlike other planets, Earth's atmosphere has the right **mix of gases** (nitrogen, oxygen, and carbon dioxide) to support life. And, it has the right amount of **air pressure** to keep temperatures even. Another important part of the atmosphere is the **ozone layer**. The ozone protects life on Earth by absorbing harmful ultraviolet radiation from the sun. So, Earth's atmosphere lets in just the right amount of warmth and sunlight to allow organisms to live and grow.

Earth also has **water** in all its forms – solid, liquid, and gas. The oceans are vast bodies of liquid water, and many tons of water evaporate from the oceans every day. This constant evaporation helps to drive the water cycle. Precipitation showers plant life and fills lakes, rivers, and streams with fresh water. No organism on Earth could survive without water.

1. List three characteristics of Earth as a planet that make it possible for there to be life on Earth.

   _____

   _____

2. Air pressure helps to keep _____.

   A) temperatures even
   B) clouds above ground
   C) airplanes from crashing
   D) none of these

3. The ozone protects life on Earth by absorbing _____.

   the water cycle   ultraviolet rays   oxygen   water vapor

4. Which of the following is an **abiotic** factor that both supports life and causes change in an environment?

   decomposers   the water cycle   predators   bacteria

5. The part of Earth between the crust and the core is the _____.

   hydrosphere   magma   mantle   loam

6. Producers use water, light, and carbon dioxide to make food in a process called _____.

7. A plant or animal that no longer exists anywhere is _____.

   extinct   scarce   exceptional   secluded

8 – 10. Use the Venn diagram to compare and contrast two kinds of animals. List as many differences and similarities between whales and sharks as you can think of.

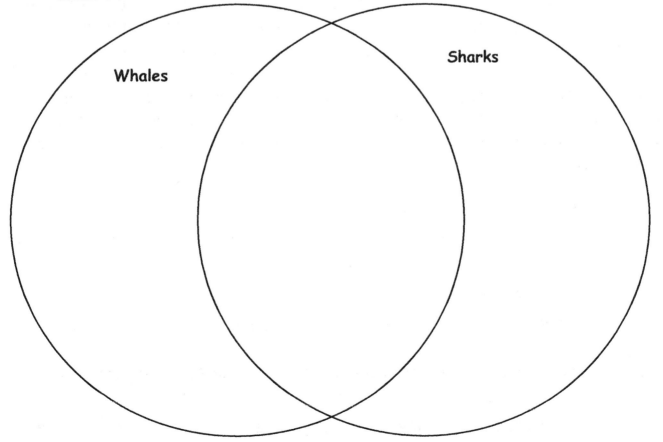

Lesson #50

The Earth's Atmosphere

Air is all around us although we can't see it, taste it, touch it, smell it, or hear it. **Air** is a mixture of nitrogen, oxygen, and other gases, along with dust particles and water vapor. These gases are tiny particles that move quickly and constantly. Gravity is pulling all of these particles toward Earth. The air surrounds Earth in layers that press on top of each other, creating air pressure. So the **air pressure** (weight of the atmosphere pressing toward Earth's core) is greater near the surface of the Earth than it is at the top of a high mountain. The air that surrounds Earth is called the **atmosphere**. Sometimes the atmosphere is called a blanket of air surrounding the Earth. This is a good way to describe it because the atmosphere keeps the Earth warm enough for Earth's living things to survive. The atmosphere is made up of many layers: **troposphere, stratosphere, mesosphere, thermosphere,** and **exosphere.**

The **troposphere** is closest to Earth's surface, and it covers Earth's entire surface including high mountains. The troposphere is a thin layer, but it contains 90% of all the gases in the entire atmosphere. The upper layers of atmosphere are resting on top of the troposphere. The weight of the upper layers presses down, making the troposphere the layer with the greatest **density**. That means there is more oxygen in the troposphere, and that is why animals are able to live on Earth. Most weather occurs within the troposphere, and all life that we know of lives within the troposphere.

The **stratosphere** is the next layer. The stratosphere contains most of the Earth's ozone (a layer of gas that absorbs the ultraviolet rays of the sun). Air in the stratosphere is thin, and it is very dry, so clouds are uncommon. The layers above the stratosphere continue to get thinner and thinner. The outermost layer – called the **exosphere** – extends indefinitely out into space.

1. Where is air pressure the greatest?

    mesosphere   stratosphere   thermosphere   troposphere

2. Most weather occurs in which layer of the atmosphere?

    thermosphere   stratosphere   mesosphere   troposphere

3. Amoeba, euglena, and paramecium belong to the same kingdom because they are all _____.

    single-celled   tiny   many-celled   hard to pronounce

Simple Solutions® Science   Level 5

4. What is the hydrosphere? _____

5. Why is the ozone important to living organisms on Earth?

   _____

6. What is air?

   gases   dust particles   water vapor   a mix of all of these

7. Which simple machine is shown in the illustration?

   _____

8. Melting and cooling are processes that form _____.

   igneous rock   sedimentary rock   minerals

9. Which scientific instrument would you use to measure length?

   balance   tape measure   thermometer   spring scale

10. Which are decomposers?

    cactus   earthworms   mushrooms   bacteria   chickens

Thermosphere

Mesosphere

Stratosphere
Troposphere

Density Think of each layer of the atmosphere as a fluffy cushion. If you stacked all of the cushions, one on top of another, the bottom layers would flatten a little from the weight of all the others. Even if all of the cushions had the same amount of stuffing, the bottom cushions would have the highest **density**. That's how it is with the Earth's atmosphere. The weight of the upper layers pressing down increases the density of the bottom layers. The troposphere is most dense, while the exosphere is least dense.

Lesson #51

1. Why is the troposphere the densest layer of the atmosphere?

    A) The troposphere is the highest layer.
    B) The weight of all the upper layers presses down on the troposphere.
    C) The upper layers have more gas particles than the troposphere.
    D) The upper layers have more dust particles than the troposphere.

2. The air pressure in the atmosphere is greater _____.

    A) the higher up you go
    B) the closer you are to Earth's surface
    C) at the top of a mountain
    D) near the moon

3. Where is the atmosphere the least dense?

    troposphere   stratosphere   mesosphere   thermosphere

4. What is formed through compacting and cementation of bits and pieces of broken down rock?

    A) igneous rock
    B) sedimentary rock
    C) metamorphic rock
    D) minerals

5. This diagram shows layers in the Earth's crust. In which layer would scientists find the oldest fossils?

    _____

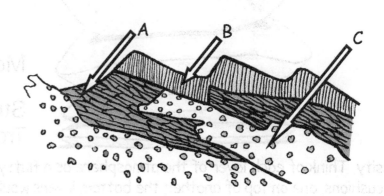

6. It seems that Earth is the only planet that can support life as we know it. Which of these is a characteristic of Earth that allows it to support life?

   A) Earth has a moon
   B) the Earth is billions of years old
   C) Earth's distance from the sun
   D) there are mountains on Earth

7. Plants use photosynthesis to help them survive and grow. What do plants need that they produce during photosynthesis?

   sunlight   carbon dioxide   food   water

8. How does a wheel-and-axle work?

   A) The wheel and axle both turn together.
   B) The wheel turns, but the axle does not.
   C) The axle turns, but the wheel does not.
   D) Neither the wheel nor the axle turns.

9. Which of these is not part of the water cycle?

   metamorphosis   precipitation   condensation   evaporation

10. How do these terms go together? Write your explanation below.

    population   carrying capacity

    _____
    _____
    _____
    _____
    _____

Lesson #52

Trish's Experiment

Trish wondered if salted water would boil faster than unsalted water. She filled a pan with one liter of plain water and heated it. Trish recorded how much time it took for the water to boil. Then Trish emptied the pan and cooled it to room temperature. She filled the pan with one liter of water and two tablespoons of salt. Using the same heat setting, Trish heated the water until it boiled, and she recorded the time.

1. What is the control in Trish's experiment? (Remember, the control is the thing that does not receive the experimental treatment.)

    heat   salted water   a timer   plain water

2. What is the **independent variable** (the thing that changes) in Trish's experiment?

    boiling   salt   heat   time

3. What is a constant in Trish's experiment?

    level of heat   amount of water   type of pan   all of these

4. List the five layers of Earth's atmosphere.

    _____ _____ _____

    _____ _____

5. An ecosystem disposes of its own waste by constantly recycling organisms and non-living materials. What may happen to an ecosystem that is not able to dispose of waste?

    A) The ecosystem may not be able to support the populations living there.

    B) Organisms may move out of the area or die.

    C) Air and water may become contaminated.

    D) All of the above

Simple Solutions© Science   Level 5

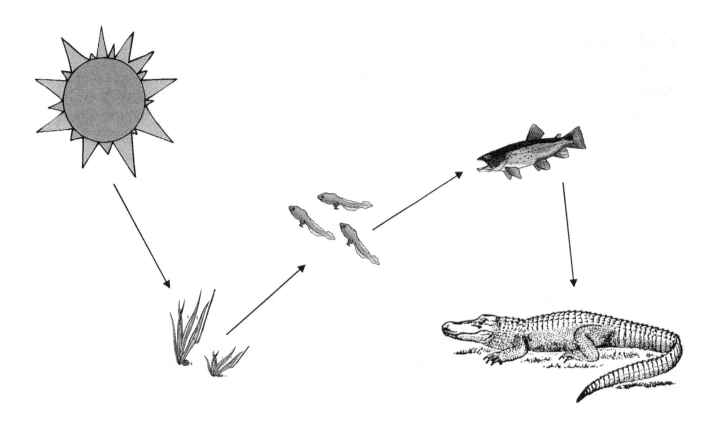

Use the food chain illustration and the words below to complete the last five items. Some of the terms will not be used.

    producers   scavenger   tertiary   secondary

    omnivore   primary   energy

6. The illustration shows that all organisms get _____ from the sun.

7. In the example above, algae are _____.

8. The tadpoles are _____ consumers.

9. The fish is a _____ consumer.

10. The alligator is a _____ consumer.

Lesson #53

What causes winds?

Air flows from an area of higher pressure to an area of lower pressure. Think about opening a can of soda; once you pop the top, pressure is released, and it immediately moves out of the can and into an area of lower pressure. Of course, the air comes out of the can quickly because there is a high amount of pressure in the can. Air will move faster when the pressure is higher and slower when the pressure is lower. If there is a big difference between the air pressures in two areas, the winds will be high. If there is very little difference, winds will be low.

Temperature is one of the factors that cause wind. Cool air is denser than warm air, so cool air sinks and warm air rises. Since the poles of the Earth are covered in ice, those areas are always cold. And the equator is always warm because it receives constant exposure to the sun. So, the cool dense air above the north and south poles sinks and moves toward the equator. Warm air above the equator rises and moves toward the north and south poles. All this movement creates a constant flow of air known as **prevailing winds**. Prevailing winds regularly blow across particular regions.

Underline the word that correctly completes each sentence.

1. Warm air (rises / sinks); cold air (rises / sinks).

2. Winds will be high if the difference in air pressure is (slight / great).

3. Air flows from _____.
   A) an area of higher pressure to an area of lower pressure
   B) an area of lower pressure to an area of higher pressure

4. Most weather occurs in which layer of the atmosphere?
   stratosphere   thermosphere   troposphere   mesosphere

5. Which is the hardest mineral?   talc   diamond   quartz   gypsum

6. How do scientists test the hardness of a mineral?
   dissolving test   scratch test   streak test   acid test

7. Name three processes that shape the surface of the Earth. (See Lesson #27.)
   _____ _____ _____

Use the diagram of a food web to complete the following items.

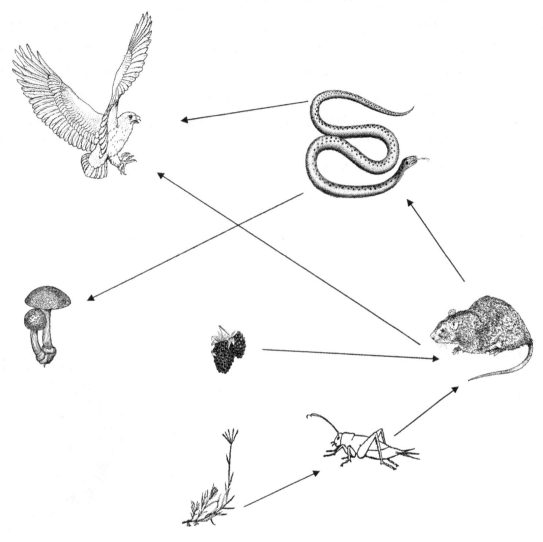

8. What will most likely happen if the population of snakes decreases?
   A) The population of mice will increase.
   B) The population of crickets will increase.
   C) The population of berry bushes will increase.

9. Which animal is an omnivore?   cricket   mouse   snake   hawk

10. Which item represents decomposers?

    mushroom   grass   cricket   snake

Lesson #54

The Meteorologist's Tools

You may not be able to see the elements that cause weather – wind, temperature, humidity, and air pressure. But you can get an idea of wind speed and direction, the amount of moisture in the air, how hot or cold it is, and the level of air pressure. Simple tools like a **wind sock** or **weather vane** show the direction of the wind. And wind speed can be measured with an **anemometer**. Measuring wind speed and direction helps meteorologists to predict the weather. **Meteorology** is the study of the atmosphere and weather conditions, and a **meteorologist** is a scientist who studies weather.

What are other instruments that measure weather? A **thermometer** is one; it measures air temperature. Meteorologists use a **barometer** to keep track of air pressure. Measuring air pressure helps to predict changes in weather. A **hygrometer** measures the **humidity** level or how much moisture is in the air.

Look in the *Help Pages* for a more complete description of tools that measure weather.

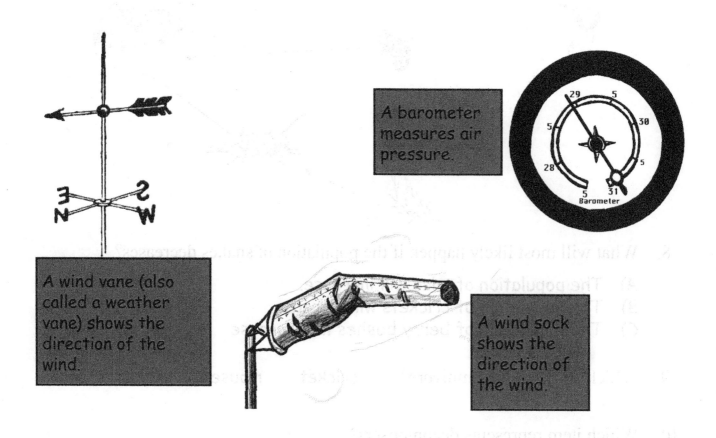

A wind vane (also called a weather vane) shows the direction of the wind.

A barometer measures air pressure.

A wind sock shows the direction of the wind.

Across

2. a measure of the amount of moisture in the air
4. the study of the atmosphere and weather conditions
6. measures air temperature
7. protects life on Earth from sun's ultraviolet rays
8. winds that regularly blow across particular regions
9. measures humidity level
10. weight of the atmosphere pressing toward Earth's core (two words)

Down

1. air that surrounds the Earth
3. shows direction of the wind (two words)
5. measures air pressure

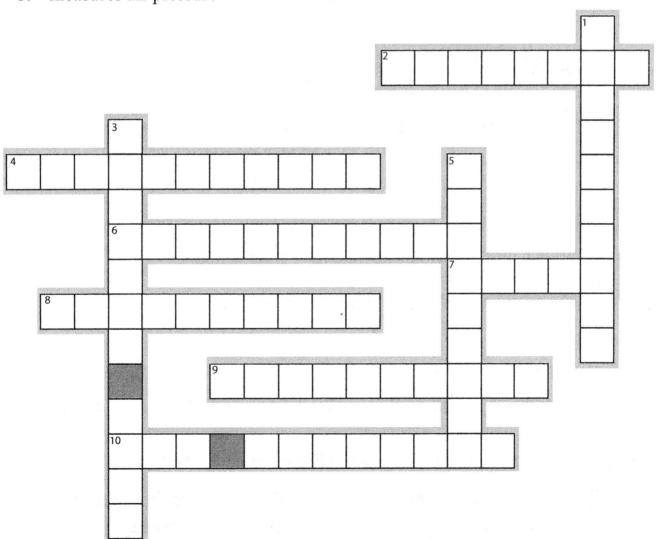

Lesson #55

The Water Cycle and Weather

You have learned that water is constantly moving and changing from one phase to another in the water cycle. Of course, the water cycle has a lot to do with weather. Water vaporizes (evaporates) becoming water vapor, and water vapor is part of the mix that we call air. The amount of water vapor that is in the air is called the **humidity** level. When there is more water vapor in the air, humidity is greater. Just like temperature, humidity changes throughout the day, from day to day, and from season to season. You may have noticed that warm summer days can be very humid (muggy), and cold winter days can be very dry. Warmer air can hold much more moisture than cooler air.

Keep in mind that warm air rises. As air rises, it expands and cools; then water vapor in the air begins to condense, forming clouds. As more and more water condenses, water droplets form within clouds, and these heavy droplets fall back to Earth's surface as precipitation. If the air is cold enough, water vapor will freeze instead of condensing. Snow, sleet, or hail falls when water vapor turns to ice.

A **hygrometer** is one of the instruments that meteorologists use to measure weather. A hygrometer measures humidity in the atmosphere.

1. Often there are heavy rains after very hot, humid days. Why is this?

    A) Warm air can hold more moisture than cool air.
    B) Warm air rises; then it expands.
    C) As the air cools, water vapor condenses, forming clouds.
    D) all of the above

2. About how long does it take for the moon to go through its phases?

    21 days　　　365 days　　　30 days　　　7 days　　　24 hours

3. Air pressure in the troposphere allows life on Earth by keeping _____.

    A) clouds above ground　　　C) airplanes from crashing
    B) temperatures even　　　　D) none of these

Simple Solutions© Science Level 5

4. _____ is a measure of the amount of water vapor in the air.

5. What does the illustration below show?

   A) a frog family
   B) the stages of metamorphosis
   C) the stages of migration
   D) a variety of vertebrate species

6. List three characteristics of planet Earth as a planet that make it possible for there to be life on Earth.

   _____

   _____

7. Draw a line through any of the following that name abiotic factors.

   plant growth   earthquake   drought   photosynthesis

8. Air flows from an area of (higher / lower) pressure to an area of (higher / lower) pressure.

9. Which organism belongs to the same kingdom as a hamster?

   mushroom   bacteria   wheat plant   dolphin

10. A _____ is a bar that pivots on a fulcrum to lift or move heavy loads.

    lever   pulley
    wheel-and-axle   screw
    inclined plane   wedge

    fulcrum

Lesson #56

Study this diagram of the rock cycle; then answer the first three questions.

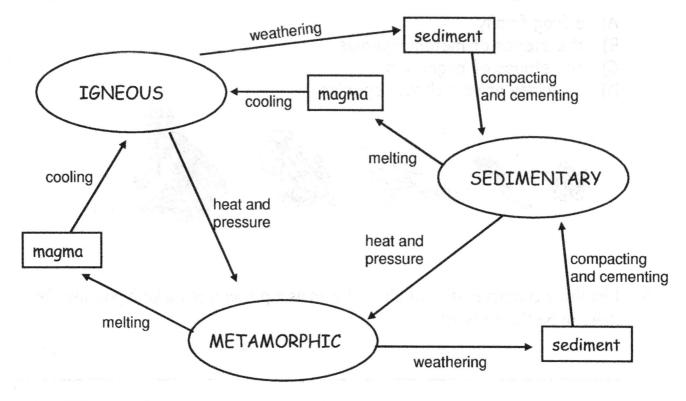

1. What are the processes that transform igneous and metamorphic rock into sedimentary rock?

    A) melting and cooling
    B) extreme heat and melting
    C) cooling and pressure
    D) weathering and compacting

2. How is igneous rock formed?

    weathering   cooling   cementing   settling

3. What processes form metamorphic rock?

    erosion & settling   heat & pressure   compacting & cementation

4. How does the ozone protect life on Earth?

    A) by producing oxygen
    B) by absorbing water vapor
    C) by increasing the water cycle
    D) by absorbing ultraviolet rays

5. What do the three animals have in common?

   A) They breathe with lungs.
   B) They live in the ocean.
   C) They are omnivores.
   D) They are invertebrates.

6. What does *inconclusive* mean?

   failed   uncertain   wrong   positive

7. What does *replicate an experiment* mean?

   try   test   repeat   explain

8. All of the waters of the Earth are known as the _____.

   environment   atmosphere   lithosphere   hydrosphere

9. Which kingdoms have organisms that can make their own food?

   animals   plants   fungi   bacteria   protists

10. One example of **how organisms interact with one another** in an environment is a _____.

    food web   weather event   volcanic eruption

Lesson #57

Air Masses and Fronts

Because the sun heats the Earth unevenly, it is always warm in some regions and cold in others. This is why air is constantly in motion throughout the atmosphere. Air moves in large buildups called **air masses**. These very large bodies of air can cover hundreds of thousands of square miles; yet an entire air mass maintains about the same temperature and level of humidity. Air masses are formed over various regions of the Earth, and they take on the temperature and humidity of the region in which they are formed. In the north, cold dry air masses are formed, while warm moist air masses are formed near the equator. Air masses may move quickly or they may barely move. When air masses with different conditions move into or out of an area, there is usually a change in the weather.

The place where two different air masses meet and bump into each other is called a **front**, and this is where weather usually changes. A **warm front** occurs when a warm air mass moves in to replace colder air. Warm air is less dense, so it rises above cooler air. Warm air also holds more moisture, so as it rises, the air begins to cool and water vapor condenses, forming rain clouds. A warm front usually brings precipitation in form of rain showers, sleet, or snow. A **cold front** occurs when cold air moves in to replace warmer air. Cold air forms a wedge underneath the warm air and pushes it aside. The warmer air rises, condenses, and brings rain. A cold front may bring heavy rain, thunderstorms, hail, or snow. A barely moving air mass is called a **stationary front**. Weather conditions may remain the same for several days under a stationary front.

1. Very large bodies of air that can cover thousands of square miles are called _____.

    prevailing winds   atmospheres   air masses   tornados

2. Warmer air always (rises / sinks) because it is less dense than cooler air.

3. Which holds more moisture?   warm air   cool air

4. What is one of the characteristics of Earth as a planet that makes it possible for living things to survive and grow?

    liquid water   air pollution   the mantle   the thermosphere

5. An orca is a whale that lives in the ocean. Whales are mammals and mammals breathe with lungs. How does the orca breathe?
   A) Whales have both gills and lungs.
   B) Unlike other mammals, whales do not need oxygen.
   C) Whales come to the surface of the water to breathe.
   D) An orca is a whale but not a mammal.

Match each of these mineral characteristics with its definition.

   A) streak   B) luster   C) transparency   D) hardness

6. _____ the ability of light to pass through

7. _____ a mineral quality that is determined by a scratch test

8. _____ the way light reflects off of a mineral

9. _____ the color of the mark that a mineral leaves when it is slid over a streak plate

10. This rock formation was most likely created by _____.

    evaporation   moving water   plant growth   condensation

Lesson #58

Climate

When people describe the type of weather in a particular part of the world, they are describing the **climate**. Seasons of the year, the amount of sunshine, average yearly temperature, and precipitation are all part of a region's climate. Some of the factors that determine climate are the location of the region, landforms, vegetation, prevailing winds, the region's elevation, and nearness to a body of water. Climate is not the same as weather. Weather is short-term and is described by the conditions – temperature, air pressure, wind, precipitation – at a certain time. Weather changes day by day and week by week. Climate on the other hand, is the established weather pattern of a certain area over many, many years.

1. What is one major difference between climate and weather?

    A) Weather is short-term; climate is a pattern that is established over many years.
    B) Weather involves precipitation; climate does not.
    C) Climate only has to do with the seasons of the year.
    D) There is really no difference between weather and climate.

2. What are some factors that determine a region's climate?

    _____

    _____

3. An air mass that is barely moving is called a _____.

    warm front   cold front   tropical front   stationary front

4. _____ is another word for a living thing.

5. Which pair of organisms is most closely related?

    A) turtle and seaweed
    B) yeast and mold
    C) monkey and tree
    D) elephant and purple bacteria

6. Stella formed a volcano out of clay. She used a mixture of baking soda, vinegar, and red food coloring to make the volcano "erupt." Stella was _____.

    doing an experiment   performing a demonstration

7. What natural force created the sea arch pictured here?

A) waves crashing against the rock over a very long time
B) builders with heavy machinery
C) a volcanic eruption
D) the movement of whales and very large fish

8. – 10. Explain what decomposers are and why they are important to all life on Earth. Use some of the words from the word bank.

| animals | bacteria | environment | food chain |
| fungi | nutrients | microorganisms | organic |

_____

_____

_____

_____

Lesson #59

The Name Says It All

Clouds come in many different shapes and sizes, and there are lots of different names for clouds. Clouds get their names from Latin words. Once you know the meanings of the Latin words, the name of the cloud will tell what kind of cloud it is. The three main types of clouds are **cirrus**, **cumulus**, and **stratus** (see below). A word that often goes with cloud names is **nimbus** which means *cloud* but always signals <u>rain</u>. Another word, **alto**, comes from a Latin word that means *high*. Altostratus and altocumulus clouds have names that are a bit misleading; while they appear to be high in the sky, they are really considered middle clouds. Cirrus clouds usually have the higher altitudes.

Latin Word	Meaning	Cloud Name	Description
cirrus	curl of hair	cirrus	wispy, like spider webs or feathers; high clouds
cumulus	heap or pile	cumulus	puffy, rippled or piled up like cotton balls
stratus	spread out	stratus	layered like blankets, mattresses, or waves
alto	high (middle)	altocumulus altostratus	puffy & patchy thin & uniform
nimbus	cloud (rain)	cumulonimbus nimbostratus	storm clouds dark, low layers

1. What does it mean if the word nimbus is attached to a cloud name?

   _____

   _____

2. Which of these is an inorganic material?

       plant leaf   mineral   tree bark   animal skin

Simple Solutions© Science   Level 5

Look at the cloud types pictured below and compare their names with information given in the chart. Choose the name that matches its description.

cirrostratus

cirrocumulus

stratocumulus

3. puffy and wave-like: cirrostratus   stratocumulus   cirrocumulus

4. wispy and layered: cirrostratus   stratocumulus   cirrocumulus

5. high, wispy, rippled: cirrostratus   stratocumulus   cirrocumulus

6. Name a natural resource. _____

7. Tissues work together to form _____.

        cells   organs   chloroplasts

8. Erosion is caused mainly by _____.

        water   rocks   plant roots   oil spills

9. Small bits and pieces of broken down rock are sometimes layered and cemented together making _____ rock.

        igneous   volcanic   sedimentary   polished

10. What are the basic needs of plants?

    A) food, water, and shelter   C) air, water, and sunlight
    B) soil, nutrients, and sunlight   D) protection, water, and oxygen

Simple Solutions© Science   Level 5

Lesson #60

How Clouds Are Formed

As you know, air is a mixture, and two of the components of air – water vapor and dust particles – are necessary in order for raindrops to form. Water vapor condenses around tiny particles of pollution or dust or even salt crystals (near an ocean). The condensation of water vapor creates clouds, and sometimes the water droplets in clouds bump against other droplets. They join together to make bigger droplets, and eventually the droplets are too big to stay in the clouds. Gravity pulls the heavy raindrops (or snowflakes if the temperature is low enough) toward Earth's surface.

1. Which two things are necessary in order for raindrops to form?

    water vapor   mountains   particles of dust   sunshine   wind

2. List the three main types of clouds:

    A) wispy   _____

    B) puffy   _____

    C) flat, layered   _____

3. To which **kingdom** does this group of organisms belong?

    shrub, flower, tree   _____

4. The natural process of breaking down rock into bits and pieces is ____.

    sediments   weathering   volcanoes   deposition

5. How do decomposers benefit plants?
   A) They kill weeds that rob plants of their nutrients.
   B) They put vital nutrients into the soil.
   C) They regulate the amount of water in soil.
   D) They gather sunlight needed for photosynthesis.

6. A _____ is a simple machine that has a slanted side and a sharp edge for sliding or for cutting. (See Help Pages.)

7. Water that falls from the atmosphere is called _____.

   evaporation   freezing   precipitation   sublimation

8 – 10. Read the descriptions of behaviors that animals use to survive. Complete the chart by filling in the word that names each behavior.

   mimicry   hibernation   migration   instinct

Description	Behavior
nest-building, migration, searching for food or protection, etc.; any behavior an animal knows without being taught	A)
conserving energy by going into an inactive state for long periods of time, especially during winter months	B)
imitating the look of another animal in order to avoid predators	C)
traveling from one place to another and back again in order to find food and shelter	D)

Lesson #61

Physical States of Matter

Everything around you and everything inside of you is made of matter. **Matter** is *anything that has mass and volume*. All matter is made up of tiny particles that can only be seen through a powerful microscope. These particles have space between them and they are constantly moving. Matter is in physical states (or phases). There are two things that determine the physical state of matter: how closely packed its particles are and how fast they are moving. Each physical state has its own characteristics or properties.

Solids, liquids, and gases are three *physical states of matter* (also called phases of matter). Usually, you can tell the state or phase of matter by observing it. For example, a paper clip, a brick, and a pencil are solids. Milk, rainwater, lemonade, and shampoo are liquids. Some examples of gases are oxygen, steam (water vapor), and carbon dioxide which we exhale when we breathe. Many gases are invisible.

1. Matter is anything that has _____ and _____.

2. The three states of matter are _____, _____, and _____.

3. Where will air pressure be greatest?

    at the top of Mount Everest   at sea level

4. A(n) _____ hatches in water and can live on land when it is grown.

    reptile   bird   amphibian   mammal

5. Which layer of Earth is thinnest?

    crust   mantle   outer core   inner core

6. What does a plant cell have that enables the plant to make its own food?

   cell membrane   cytoplasm   chloroplasts   none of these

7. Which of the following are abiotic factors that can cause change in an environment?

   volcanic eruption   seeds sprouting   cell division   earthquake

8 – 10. Energy is transferred among these organisms in an aquatic food chain. Use words and/or drawings to explain the transfer of energy; put your answer in the box.

Alligator   Heron   Salamander   Insect

Simple Solutions© Science — Level 5

Lesson #62

Mass and Volume

Matter is anything that has **mass** and **volume**. **Mass** is *the amount of matter in an object* or the amount of "stuff" that an object is made of. The mass of an object <u>does</u> <u>not</u> <u>change</u>. A cup of milk has a certain mass. You can divide the cup of milk into three containers, but if you add up the mass held in all three containers, it will be exactly the same as the mass in the original cupful. **Volume** is *the amount of space matter takes up*. Unlike mass, the volume of matter may change. As the particles within an object move closer together (or further apart), the amount of space the object takes up will change. In all matter, mass is definite (it never changes), but volume is indefinite (it may change).

Mass is not the same as weight. An object's **weight** is a measure of the force of gravity on the object – how gravity pulls on the object. Your weight on earth would be greater than your weight on the moon because the force of gravity is much greater on Earth than it is on the moon. But, as we said above, your mass would not change; it is the same wherever you are.

1. _____ is the amount of matter in an object. _____ is the amount of space matter takes up.

2. What is one difference between mass and weight?

    A) An object's mass never changes; its weight may measure differently depending on the force of gravity.
    B) Mass is greater than weight.
    C) Weight can be measured; mass cannot.
    D) Objects only have weight on Earth; objects have mass anywhere.

3. The troposphere contains _____ of the gases in the atmosphere.

    25%   90%   10%   100%

4. The place where two air masses meet is called a _____.

    A) system
    B) hurricane
    C) front
    D) local wind

Scientists use a balance to measure the mass of an object.

5. Which process is necessary to the formation of clouds?

   weathering   precipitation   condensation   deposition

6. Which word part lets you know that a cloud is likely to bring heavy rain?

   alto   stratus   nimbus   cirrus

7. Fish, birds, amphibians, reptiles, and mammals are all _____.

   vertebrates   invertebrates

8. Brian noticed mushrooms growing near the edge of his lawn. What are the mushrooms a sign of?

   A) drought
   B) a type of flowering plant
   C) decaying organic matter
   D) flooding

9. What is a hypothesis?

   a thought   a proven theory   an educated guess   none of these

10. Below are diagrams of the Earth's layers. What is one explanation for the gap in the middle of each diagram?

    A) Magma is pushing up and breaking through other layers.
    B) An earthquake has caused a crack and an opening in the crust.
    C) Either of the above is a possible explanation.

Lesson #63

Solids

The particles that make up a **solid** are very close together and do not move around very much at all; the particles of a solid move very slowly. This makes solids rigid, and that is why solids keep their shape. Remember, having **volume** means the object takes up space. For solids, the volume doesn't change very much at all. In fact, it changes so little, that we can hardly notice it. For this reason, we say that the volume of a solid is definite (unchanging). A solid keeps its volume and shape because it contains tightly packed particles that are barely moving.

A solid keeps its shape and volume whether it is inside or outside of a container.

1. A solid does not change its shape or volume because the particles in a solid are _____.

    packed closely together   held tightly   both of these

2. The most common states of matter are _____, _____, and _____.

3. Which affects weather?

    temperature   air pressure   wind speed   all of these

4. Insects and some other invertebrates have hard outer shells called ___.

    vertebra   exoskeletons   gills   scales

5. Which pair of organisms is most closely related?

    whale/snake   mosquito/grass   mushroom/toad   shark/moss

Simple Solutions© Science Level 5

6. What is one main difference between an experiment and a demonstration?

   A) The outcome of a demonstration is known ahead of time, but the outcome of an experiment is not known ahead of time.
   B) An experiment proves that a hypothesis is correct; a demonstration does not.
   C) An experiment must be performed in a lab; a demonstration can take place outside of a lab.
   D) There is no difference between an experiment and a demonstration.

7. Which simple machine is on the tines of a fork?

   screw   wedge   inclined plane   pulley

8. Which of these describes a primary consumer?

   omnivore   carnivore   producer   herbivore

9 – 10. How are these phrases similar? How are they different? Write your explanation below.

   metamorphic rock   stages of metamorphosis

   _____

   _____

   _____

   _____

   _____

Lesson #64

Liquids

The particles that make up a **liquid** have more space between them and move faster than the particles of a solid. Particles in liquids are able to move around a little more. That makes liquids able to flow. Unlike solids, liquids do not have a definite shape – they adapt to the shape of their container.

Look at the two containers below. Can you see spaces around the pieces of paper? The paper is in the container, but each piece has its own shape. Imagine what would happen if you emptied out both containers onto a table. You could pick up the wads of paper. But without the glass, the milk would spread out into a puddle or run off the edge of the table. The amount of **mass** in the paper and the amount of **mass** in the milk remains the same whether the paper or milk is in its container or not. The shape can change, but the mass never changes!

Just as with solids, the volume of a liquid doesn't change very much at all. Both solids and liquids have a definite volume. That means the volume does not change.

Particles in a solid are tightly packed, so a solid keeps it shape inside or outside of a container.

Particles in a liquid are loose, so a liquid flows and takes the shape of its container.

1. How is a solid like a liquid?

    A) Both have a definite shape and volume.
    B) Both have a definite volume.
    C) Both have a definite shape.
    D) A solid is not like a liquid in any way.

2. Liquids are able to flow because the particles _____.

    A) have space between them
    B) move freely and can slide over and around each other
    C) Both A and B

3. Which of these is made of particles that are not moving at all?

    solid   liquid   gas   none of these

4. When you put a solid into a jar, it will _____.

    change its shape   change its size   keep its shape and volume

5. Which two of these words may describe a tertiary or third-level consumer?

    herbivore   carnivore   producer   omnivore

6. Which type of cloud looks like wisps of curly hair or spider webs?

    _____

7. Which type of rock is formed when volcanoes erupt and molten rock cools?

    A) magma   C) igneous
    B) magnetite   D) sand

8. A(n) _____ is an animal with hair or fur that feeds its young with milk.

    reptile   mammal   amphibian   avian

9. Which simple machine is pictured?

    _____

10. What is the job of this simple machine?

    _____

    _____

    _____

Lesson #65

Gases

The particles in a gas have a lot of space between them, and they move freely. So gas particles can break apart from one another. Gas particles will continue to move farther and farther apart unless something stops them. The volume of a gas will increase until it completely fills its container. That's why we say <u>a gas has neither a definite shape nor a definite volume.</u>

For example, a balloon can hold a certain amount of helium. If you open the balloon, the helium will spread out into the room. If a window is open, the helium will continue to expand out into open space.

| A solid has a definite shape and volume. It keeps its shape and volume whether it is inside or outside a container. | A liquid has a definite volume but not a definite shape. A liquid changes its shape according to the container. | A gas does not have a definite shape or volume. A gas takes the shape of its container. As the gas leaves its container, its particles get farther and farther apart. |

Fill in the blanks; use the words **solid**, **liquid**, or **gas**.

1. A _____ takes the shape of its container, but it does not change in volume.

2. A _____ has a definite volume, and it keeps its shape, no matter what container it is in.

3. A _____ has no definite shape and no definite volume. It expands to fill all the space available.

4. A balance is used to measure _____.

        weight   mass   both   neither

Simple Solutions© Science

Level 5

5 – 10. Complete this crossword puzzle using the hints below.

Across
1. level of consumer that eats animals which eat plants
3. level of consumer that eats animals which eat other animals
5. breaks down dead plants and animals
6. able to make its own food; first in the food chain

Down
2. eats both plants and animals
4. level of consumer that eats producers only

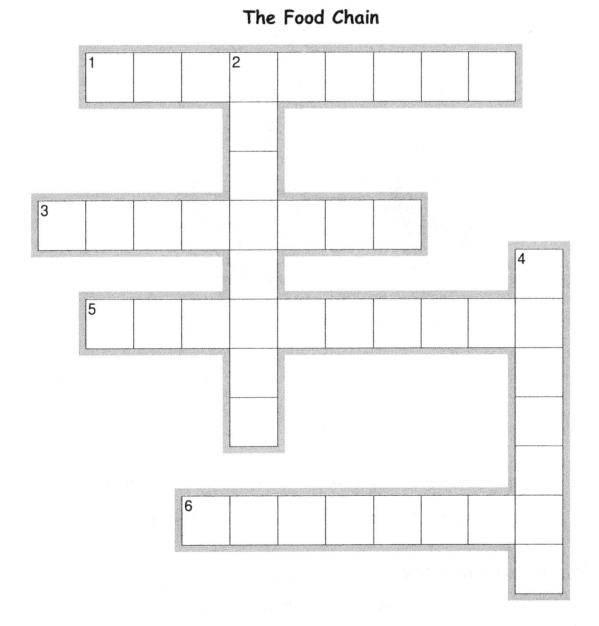

The Food Chain

Lesson #66

Phases of Matter

States of matter are also called **phases of matter**. Matter may change from one physical phase to another when there is a change in **temperature**. For example, when enough heat is applied to a solid, it will **melt** and become liquid. More heat will make the liquid **boil**. (The melting point of a substance will always be lower than its boiling point.) During boiling, the liquid becomes vapor (gas). When enough heat is removed from a liquid, freezing or **solidification** occurs. The liquid moves into its solid phase.

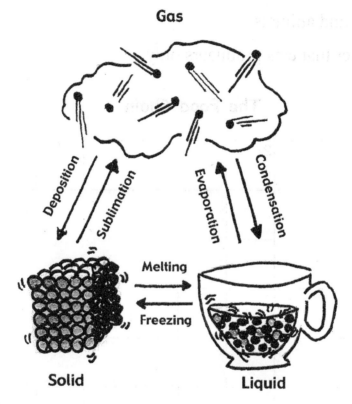

1. What are the three states of matter?

   _____   _____   _____

2. What is another word for *freezing*?   _____

3. The melting point of a substance will always be lower than its _____.

   freezing point   boiling point   neither   both

4. Which has the most freely moving molecules?

   solid   liquid   gas

5. Water turns to vapor in a process called _____.

6. Which has the slowest moving particles?   solid   liquid   gas

If you pour one cup of lemonade into a glass, it will take up a certain amount of space in the glass. If you pour that same cup of lemonade from the glass into a jar, it will take up the same amount of space, even though its *shape* will be different. Another way to say it is that the lemonade has the same **volume** whether it is in the glass or the jar. Liquids change their shape because they take on the shape of the container. If you pour the cupful of lemonade into a cone, it will take the shape of the cone. However, the volume of the lemonade (one cup) will be the same in all three containers.

7. In the example described in the paragraph above, what is the volume of the lemonade?

   A) one cup
   B) the same as the mass
   C) one glass full
   D) The volume would be different in each container.

8. Why are minerals called **inorganic**?

   A) Minerals are not and never were living.
   B) Minerals come in many different colors.
   C) Minerals have various properties.
   D) Minerals contain bits of dead plant material.

9. Which illustration shows an instrument that measures wind speed?

   A)    B)    C)   D)
   Barometer

10. Which simple machine makes it possible to fasten the lid onto a Mason jar?

    A) wedge   D) wheel-and-axle
    B) pulley   E) lever
    C) inclined plane   F) screw

Lesson #67

Some of the Properties of Matter

Matter has **properties**: how it looks, how it acts, and how it feels. Some of the properties of matter are **physical state, melting point, freezing point, condensation point**, and **density**. You already know that matter has mass (the amount of stuff in a thing). You know that mass can be measured using a balance. You know that the physical state of matter can change – solids can become liquids; liquids can become gases, and so on. Every substance has a **melting/freezing point**. This is the temperature at which a substance changes from solid to liquid and vice versa. For example, liquid water freezes at 32°F (0°C); solid water (ice) melts at the same temperature — 32°F is waters freezing point and it is also ice's melting point.

Each substance also has a **boiling/condensation point** – the temperature at which the substance changes from liquid to gas and vice versa. For water, the boiling/condensation point is 212°F (100°C).

1. What are some of the properties of matter?

   _____

Write **M** for melting point, **F** for freezing point, or **C** for condensation point.

2. _____ temperature at which a solid will become a liquid

3. _____ temperature at which a liquid will become a solid

4. _____ temperature at which a gas will become a liquid

5. If you took a string of licorice and cut it up into little pieces, how would that affect the **mass** of the licorice?

   A) The mass of the licorice would be greater.
   B) The mass of the licorice would be less.
   C) The mass of the licorice would not be changed.
   D) The mass would be impossible to measure.

Simple Solutions© Science   Level 5

6. The photo shows that a _____ has no definite shape.

   gas   liquid   solid   all three

Hypothesis: Microwaveable popcorn will pop best if stored at room temperature.

Materials: four bags of microwave popcorn (same brand and date), microwave, and storage areas (refrigerator, freezer, heated area, other)

Procedure:

- Store bags of microwave popcorn for one week in four different locations. Put one bag in the refrigerator, put another in the freezer, place a third bag in an area at room temperature (such as a kitchen cupboard), and put the fourth bag in a heated area (such as near a radiator or under a heat lamp).

- Pop the bags of popcorn in a microwave, one bag at a time, according to package directions. Use the same number of minutes and the same heat setting for each bag.

- Pour the contents of each bag onto a table or tray and separate the popped kernels from the un-popped kernels. Count the un-popped kernels and record this information in the data chart.

7. The independent variable being tested in the experiment is _____.

   popcorn   number of kernels   storage temperature   brand

8. When the tester is popping the popcorn, why is it necessary to use the same number of minutes and the same heat setting for each bag?

   A) to isolate the variable
   B) so the popcorn won't burn
   C) to keep the microwave from overheating
   D) to make sure that some of the kernels do not pop

9. Organisms and the environment they live in make up a(n) _____.

   population   ecosystem   hydrosphere   food web

10. The smallest part of a living thing is a _____.

135

Lesson #68

Properties of Matter continued: Phase Changes

Matter may change from one physical phase to another when there is a change in **temperature**. Water is a great example of how a substance changes from one state or phase to another. You have seen water in all its phases. The temperature at which a substance will boil is called its boiling point. The **boiling point** of water is 100° C or 212°F. Once water starts to boil, you will see steam (water vapor) rising from it. Water turns to vapor (a gas) in a process called **evaporation** (or **vaporization**).

The temperature at which a substance will freeze (or solidify) is called its freezing point. The **freezing point** of water is 0° C or 32° F. When water changes to ice (solid), the process is called **freezing**. When heat is applied to ice (solid), it changes to water (liquid) in the process called **melting**.

Vapor (gas) changes to water (liquid). This process is called **condensation**. A solid can also change directly to a gas; this process is called **sublimation**. Sublimation occurs when ice or snow changes directly to vapor. The solid "sublimes." That means it goes from solid to gas without becoming a liquid first.

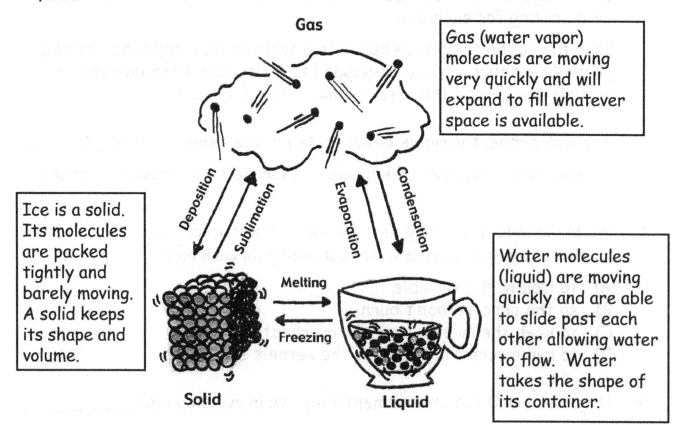

1. What are the three states or phases of matter?

_____   _____   _____

Simple Solutions© Science   Level 5

2. What is another word for *freeze*?

   _____

3. The melting point of a substance will always be lower than its _____.

       freezing point   boiling point   neither   both

4. Which has the fastest moving molecules?   solid   liquid   gas

5. Water turns to vapor in a process called _____.

6. A solid can also change directly to a gas in a process called _____.

7. Which pair of organisms is most closely related?

       rat/donkey   amoeba/ant   zebra/fern   dolphin/moss

Mr. Roberts held up a fully inflated balloon and a long thin needle. He inserted the needle all the way through the balloon without popping it. Mr. Roberts explained that balloons are made of a polymer that is porous (full of tiny spaces).

8. Mr. Roberts was _____.
   - A) conducting an experiment
   - B) performing a demonstration
   - C) teaching magic tricks

9. The innermost part of the Earth is its _____.

       mantle   crust   core   atmosphere

10. Which organism is most likely to be at the bottom of a food chain?

        algae   crocodile   rabbit   grasshopper   skunk

Lesson #69

Density

Here is a demonstration you can try: Measure ¼ cup of maple syrup and put it in a beaker; then measure ¼ cup of liquid soap and carefully pour it in on top of the syrup. The soap will float on top of the syrup. Next, measure ¼ cup of water and carefully pour it into the beaker; the water will float on top of the soap. Finally, add ¼ cup cooking oil. You will see a layered column of liquids. Why? Well, ¼ cup of syrup is denser than ¼ cup of water; ¼ cup of liquid soap is denser than ¼ cup of water. Cooking oil has the lowest density, so cooking oil would float on top of the syrup, the soap, and the water.

Matter has density. Two different kinds of matter that have the *same volume* may each have a completely *different mass* (or vice versa). Therefore, one has a greater density than the other. **Density** is a measure of how closely molecules are packed in a given space. How dense something is depends upon both its mass and its volume. In the example, the syrup is denser than the water, and the water is denser than the oil, even though all the liquids have the same volume. The substances with greater density will sink below those with lower density. The substances with lower density will rise above denser substances. Density is just one of the physical properties of matter. Physical state, boiling point, and melting point are some of the others.

1. _____ is the amount of matter in an object.

2. Which would have a greater density, a box full of foam peanuts or box full of pennies?

   _____

3. Water will become vapor when it reaches its _____ point.

       freezing   boiling   melting   solidification

4. Matter may change from one physical phase to another when there is a change in _____.

       mass   color   temperature   application

5. What is humus?

   decaying organic material   rock particles   mineral deposits

6. Which of the following is an abiotic factor that may cause change in an environment?

   scavengers   predators   plant roots   earthquakes

7. Magma is melted rock. Some examples of other substances that melt are butter, chocolate, wax, and ice cubes. What could cause a substance that has melted to solidify?

   thawing   heating   cooling   settling

8. The imprint or remains of things that lived long ago are called _____.

   photographs   fossils   remnants   artifacts

9. What do plant cells have that allows the plant to stand firm? (Animal cells do not have this.)

   _____

10. What does *inorganic* mean?

    hard   living   non-living   transparent

    High Density   **Low Density**

    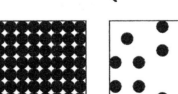 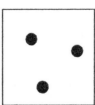

Density is a measure of how closely molecules are packed in a given amount of space. When the volume is the same, the matter with the greatest number of molecules has the greatest density.

Lesson #70

Mixtures

A **mixture** is a combination of two or more substances. You are making a mixture when you make trail mix; it's a mixture of cereal, nuts, and dried fruit. Almost anything you can think of in nature is a mixture. Sand, ocean water, air, glaciers, and wood are a few examples. As you know, soil is a mixture – it contains bits of rock, air, water, and organic material. Manufactured materials are also mixtures: concrete, plaster, plastics, glass, paint, and so on.

The parts of a mixture are not permanently combined. In other words, each ingredient in the mixture can be taken back out of the mixture. It would be easy to spread trail mix out on a table and separate it back into piles of nuts, fruit, and cereal. Other mixtures are not as easy to separate, but taking them apart is not impossible. The properties of each substance do not change by becoming part of a mixture. For example, mud is a mixture of dirt and water. Neither the dirt, nor the water has changed. Also, the dirt can be separated from the water by letting the water evaporate out of the mud.

1. What is a mixture?

   _____

2. Give an example of a mixture.

   _____

A mixture is a blend of any two or more substances. The parts of a mixture are held together by physical forces.

3. Which is true?

   A) A mixture can be separated.
   B) A mixture is a permanent combination of substances.

4. Marty noticed that every time he took salad dressing out of the refrigerator, the oil was floating on top of the vinegar. Why?

   A) Cold air causes liquids to separate.
   B) The makers of the salad dressing want you to shake it up.
   C) Oil is denser than vinegar.
   D) Vinegar is denser than oil.

5. Phillip placed an egg in a beaker full of tap water; the egg sank to the bottom. Then he removed the egg and dissolved 8 tablespoons of salt in the tap water. When Phillip put the egg in the salt water solution, the egg floated! What is the best explanation for why this happened?

   A) The egg was cracked when Phillip put it in the beaker the second time.
   B) The egg was raw the first time and hard boiled the second time.
   C) Salt water has a lower density than plain tap water.
   D) Salt water has a higher density than plain tap water.

6. Liquids flow, but the particles in a liquid are bound to each other and will stay connected. Particles in a gas _____.

       also stay connected   move apart easily   hardly move at all

7. True or False?

   A) _____ Melting point is higher than boiling point.

   B) _____ Boiling point is higher than melting point.

   C) _____ Melting point and boiling point are equal.

8. When should you wear safety goggles?
   A) any time you want to look cool
   B) when reading a science text
   C) when handling chemicals
   D) any time you are outdoors

9. All of the living and nonliving things interacting and affecting each other in a certain area are a(n) _____.

       migration   ecosystem   population   kingdom

10. _____ are cold-blooded vertebrates that begin life breathing with gills; later they develop lungs.

        Mammals   Birds   Reptiles   Amphibians

Simple Solutions© Science   Level 5

Lesson #71

Solutions

   A **solution** is a type of mixture, and just as in other mixtures, the molecules of a solution can be separated from one another. But in a solution, all the parts are evenly distributed (spread out). A solution is called homogeneous because the particles all look the same. A solution is still a mixture – it is a liquid dissolved in another liquid, or a solid dissolved in a liquid, or a gas dissolved in a liquid. But in a solution, the molecules are mixed evenly and all look the same. A solution is so well blended that it looks like it is just one substance. Saltwater is a solution. If you took a sampling of saltwater from the top of a glass and another sampling from the bottom of the glass, both samplings would have the <u>same distribution</u> of salt in water.

   A solution may be made by dissolving a solid into a liquid – for example, fruit punch made from a powder. When you pour a scoopful of instant drink mix into a pitcher of water and stir, you are making a solution. (Remember, a solution is a mixture in which all of the parts are <u>evenly distributed</u>).

1. A solution is a mixture in which all of the particles are _____.

   the same   mixed evenly   easy to see   impossible to separate

2. Salad dressing is a (mixture / solution) of water, oil, vinegar, and spices.

> A solution is a mixture, but not all mixtures are solutions.

3. About how long does it take for the moon to go through its phases?

   21 days   365 days   30 days   7 days   24 hours

4. A paper cutter is a combination of which two simple machines?

   A) pulley and screw
   B) lever and wedge
   C) lever and inclined plane
   D) wheel-and-axel and screw

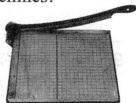

142

Simple Solutions© Science   Level 5

5. Water will change from liquid to solid when it reaches its _____ point.

   boiling   condensation   purification   freezing

6. Ellen placed an apple on a balance and found that its mass was 302 grams. Then she cut the apple into six slices and placed it back on the balance. The mass was still 302 grams. What was Ellen demonstrating?

   A) The balance needs to be adjusted.
   B) Mass does not change.
   C) Apples have no volume.
   D) An apple is a solid.

7. How do scientists classify all animals which have backbones?

   animal kingdom   vertebrates   mammals   consumers

8. Give an example of how organisms interact with each other in their environment. (See Lesson #20.)

   _____

   _____

9. A third-level consumer eats _____.

   plants only   animals   producers

10. Why does hot air rise?

    A) Hot air is less dense than cooler air.
    B) Hot air is heavier than cool air.
    C) Hot air contains a lot of helium.
    D) Gravity causes hot air to rise.

Lesson #72

1. Which pair of organisms is most closely related?

    spider/sheepdog   spider/mushroom   spider/algae

2. What of these is true?

    A) There is no difference between a mixture and a solution.
    B) A mixture can be "unmixed," but a solution can not.
    C) A solution is homogenous, and the molecules are evenly distributed.
    D) A mixture may not contain more than two substances.

A mixture is a solution if its parts are evenly distributed.

A solution may look like it is just one substance.

3. Which of these is not a simple machine?

    spring   screw   ramp   pulley   lever

Properties are like characteristics – anything that describes a substance. For example, what are the properties or characteristics of a wax candle? It has a color, texture (how it feels) and it may have a scent. A wax candle has a melting point, and there is a temperature at which the melted wax will solidify.

4. The properties of a substance are _____.

    A) anything that describes it
    B) only boiling point and melting point
    C) color and texture only
    D) fire and temperature only

144

5. Which of these is not one of the properties of a solid?

    A) has density
    B) has mass
    C) takes the shape of its container
    D) has volume

6. Fill in the word that names how matter changes from one phase or state to another. Use these words: **deposition, freezing, condensation, sublimation, evaporation, melting**.

Process	Change	Example
	solid to liquid	butter on a hot ear of corn
	solid to gas	ice cubes disappearing in the freezer
	liquid to solid	water in a pond becoming a solid skating rink
	liquid to gas	boiling water
	gas (vapor) to liquid	dew
	gas to solid	frost on windows

7. Which of the following are adaptations that animals use for survival?

    hibernation   mimicry   camouflage   migration   all of these

8. The oldest fossils will most likely be found in the _____ layers of Earth's crust.

    thickest   deepest   middle   most shallow

9. Which of these describes a primary consumer?

    omnivore   carnivore   producer   herbivore

10. What does a hygrometer measure? _____

Simple Solutions© Science   Level 5

Lesson #73

Physical Change

You already know that matter has properties (characteristics) such as color, density, melting point, and freezing point. Also, matter can change from one state (or phase) to another. A solid can become a gas; a gas can become a liquid and so on. These are physical changes. A **physical change** does not change what the substance is. If you chop up a piece of wood, it is still wood; if you freeze a gallon of water, it is still water. If you shape some clay into a sculpture, it is still clay, and if you crush a rock into a powder, the powder is still bits of the rock. Even if you cut off some of your hair, it is still hair and you are still you! Physical changes can be reversed. For example, you can blow up a balloon and then deflate it. Or, you can paint a piece of furniture; then remove the paint. You can wrinkle up a piece of fabric, and then iron it out; you can bend a wire and straighten it. During a physical change, <u>matter changes in its form but does not change in its substance</u>.

1. What are some of the properties of matter?

   _____

2. Which of these describes a **physical change**?

   A) The substance changes in form, but it is still the same substance.
   B) The substance changes completely and is made of different atoms.
   C) Some of the substance disappears completely.
   D) A totally new substance is formed.

3. Name the five kingdoms. (See Lesson #17 if you need help.)

   _____   _____

   _____   _____

   _____

4. Which are characteristics of Earth as a planet that make it possible for there to be life on Earth?

   distance from the sun   space exploration
   water cycle   carbon emissions

5. Below is a diagram of an area where paleontologists are digging for fossils. In which layer will they probably find the oldest fossils?

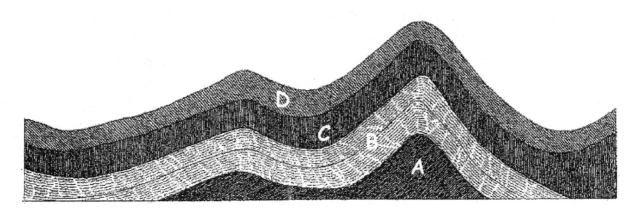

6. The atmosphere supports life on Earth by _____.

    A) providing the right mix of gases
    B) keeping temperatures even
    C) holding water vapor
    D) all of the above

7. An ecosystem disposes of its own waste by constantly recycling organic material. How does an ecosystem recycle organic waste?

    A) Decomposers do the work.
    B) A recycling center is used.
    C) Waste materials are put in landfills.
    D) Organisms move waste materials to other ecosystems.

8. About how long does it take for the moon to go through its phases?

    365 days   30 days   60 minutes   7 days   24 hours

9. Which of these is a good conductor of electricity?

    wood   steel   rubber   Styrofoam

10. Which type of cloud looks like a head of cauliflower?

    cirrus   stratus   cumulus   altostratus

Lesson #74

Chemical Change

You learned that matter can change in *form* during a physical change. Matter can also change in *substance*. During a **chemical change** (or chemical reaction), *an entirely different substance is formed.* This is because the atoms in the substance actually get rearranged. When you see rust, you are looking at the result of a chemical change. Iron combines with oxygen to form rust; however, rust is neither iron, nor oxygen. It is a different substance with its own make-up and properties. The molecules that make up iron and oxygen have been rearranged. Another example is burnt food. When a marshmallow burns, a chemical change takes place. The sugar molecules that make up the marshmallow break down into carbon and water — different substances with their own make-up and properties.

Here's another example: If you combine baking soda and vinegar, you will see bubbling, which is evidence of a chemical reaction. The result will be carbon dioxide (a gas) and water.

1. What is another term for a **chemical change**? _____

2. Which of these is an example of a **physical change** only?

    crushing   corrosion   burning   all of these

3. Which is an example of a chemical change?

    melting   freezing   burning   all of these

4. Matter is anything that _____.
   A) takes up space
   B) is made of atoms
   C) has mass
   D) all of the above

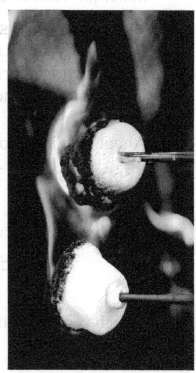

When marshmallows burn, there is a chemical change.

5. Why should you never eat or drink out of laboratory containers or utensils?
   A) Lab containers may be coated with toxic chemicals.
   B) There is no sink for washing the containers.
   C) Lab utensils and containers are made of glass.
   D) Other people may think this is rude.

6. The natural process of breaking down rock into bits and pieces is _____.
   sediments   weathering   volcanoes   deposition

7. Which type of cloud looks like a blanket covering the sky?
   cumulus   cirrus   altocumulus   stratus

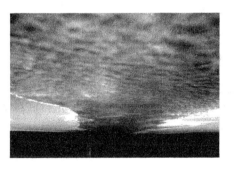

8. Which type of soil is the best for growing crops?
   clay   loam   sand   silt

9. Which layer of the atmosphere is closest to the Earth's surface?
   thermosphere   mesosphere   troposphere   none of these

10. Look at the list below. All of these organisms have something in common except for one. Which one does not belong?
    whale   fish   bacteria   blue jay   toad

    Explain why the one you chose does not belong.

    _____

    _____

Lesson #75

The Law of Conservation of Matter

You have learned that everything you see (and even what you don't see) is made up of matter, and matter is anything that has mass and volume. All of this matter can never be destroyed, and new matter is not created. Matter simply changes from one form to another during a physical change or a chemical reaction. The **Law of Conservation of Matter** states that **mass is neither created nor destroyed**. That means that everything "new" is really just stuff that was already here although it may have taken a new form. Here is an example: Let's say you take a block of ice and measure its mass using a balance. Then you allow the block of ice to completely melt, and you measure the mass of the liquid water. The measurements will be exactly equal because all of the atoms that make up the water are still there whether the water is frozen or not.

What if you completely burn something, like a sheet of paper? You know that burning creates a chemical reaction, but even chemical change does not destroy matter or create it. The chemical reaction – burning – simply causes a *change* in matter. <u>Burning rearranges the atoms into new substances</u>, like carbon and water vapor. If you could measure the mass of these substances, you would see that the mass is exactly the same as the mass of the original sheet of paper.

1. According to the Law of Conservation of Matter, when water evaporates, _____.

    A) the water molecules disappear
    B) more water is created in the form of precipitation
    C) the molecules change but are not destroyed
    D) water is removed from the universe

2. According to the Law of Conservation of Matter, mass is neither _____, nor _____.

3. A **chemical reaction** results in _____.

    A) an increase in matter
    B) the formation of a different substance
    C) no permanent change
    D) a change that can be reversed

4. _____ is the population size that an ecosystem can support without damaging itself.

   Census   Carrying capacity   Extinction   Migration

5. Where are clouds?

   in the troposphere   in the exosphere   above the stratosphere

6. If the "like" poles of two magnets are put next to each other, the poles will _____.

   repel   attract   do nothing

7. The basic units of life in all organisms are _____.

   chloroplasts   nuclei   cells   proteins

8. Animals that do not have a backbone are called _____.

   invertebrates   amphibians   mollusks   vertebrates

9. Water vapor becomes liquid water at its _____ point.

   melting   freezing   condensation   hydration

10. According to laboratory safety rules, one thing you should never do in a science lab is _____.

    A) pour chemicals down the drain
    B) take notes during a lab activity
    C) work with a partner
    D) wear latex gloves

Lesson #76

1. According to the Law of Conservation of Matter, mass is neither _____ nor _____.

2. In animals, cells work together to form _____.
       tissue   producers   nuclei   vacuoles

3. True or False?

   A) _____ A chemical change does not create new matter.

   B) _____ A physical change does not destroy matter.

   C) _____ A physical or chemical change may create more matter.

4. Why will balloons filled with helium float to the ceiling in a room?
   A) Helium is less dense than the air in the room.
   B) Latex balloons are lighter than air.
   C) The air temperature in the room is higher near the ceiling.
   D) Static electricity attracts the balloons toward the ceiling.

5. Marti visited the pond in the park near her house one very cold afternoon. The pond was a solid sheet of ice, and skaters were gliding across in every direction. What kind of change made the pond into a skating rink?
       chemical   physical   extraterrestrial   geothermal

6. Which Latin word names clouds that are likely to bring rain?
       geo   alto   nimbus   cirrus   thermal

Simple Solutions© Science Level 5

7. If an ecosystem is not able to properly dispose of waste, what will happen?
   A) The ecosystem may become polluted.
   B) The ecosystem's carrying capacity will be decreased.
   C) Some animals may die off or move out of the ecosystem.
   D) All of the above may happen.

8. From what does a decomposer get energy?

   cell division   photosynthesis   dead things   water

9. Another term for harmful pollutants is _____ waste.

   hazardous   nonrenewable   recyclable   compound

10. Complete the chart with three examples of each of the states of matter.

Solid	Liquid	Gas

Lesson #77

What is Energy

Energy is the ability to do work. What does this really mean? Another way to say it is that energy is a "doer." Energy changes and causes change. You experience energy every day in its many forms – heat, light, sound, and motion are a few of these forms. As **heat**, energy warms your home and cooks your food. **Light** energy illuminates your study area and makes plants grow. **Sound** is energy that allows you to hear your favorite tunes. **Motion** energy moves a basketball through the air.

1. What is the definition of energy?

   _____

2. Another term for "an educated guess" is _____.

       model   variable   hypothesis   constant

3. Which pair of organisms is most closely related?

   alligator/mold   iguana/toadstool   fly/moss   ivy/dogwood tree

4. Different substances are formed when atoms are rearranged in a (physical / chemical) change.

5. Which of these tells why sand is called a **mixture**?

   A) Sand is a combination of two or more substances.
   B) Sand takes a very long time to form.
   C) Sand is coarse and has some very large particles.
   D) Sand contains organic material.

Sand contains bits of rock, salt, coral, and shells.

6. The photo below shows that a _____ has neither a definite shape nor a definite volume.

   gas   liquid   solid   all three

7. Which simple machine allows skaters to move on roller blades?

   A) pulley   E) wheel-and-axle

   B) screw   D) inclined plane

   C) lever   F) wedge

8. ____ is the weight of the atmosphere pressing toward Earth's core.

   Air pressure   Ozone   Prevailing winds   Gravity

9. Read each statement. Write **T** if the statement is true or **F** if it is false.

   A) ____ Minerals have properties like color, luster, and streak.

   B) ____ Rocks are hard, but minerals are soft.

   C) ____ Minerals are inorganic.

   D) ____ Rocks contain minerals, but minerals are not rocks.

10. What is the Law of Conservation of Matter?

_____

_____

Simple Solutions© Science — Level 5

Lesson #78

Kinetic Energy

All energy is either potential or kinetic. **Kinetic** (ki net' ik) energy is energy in motion. It lights up houses, moves cars along the road, and melts the snowman in your yard. A baseball flying through the air has kinetic energy. A person running down the stairs, a wave crashing on a shoreline, and a bird soaring across the sky all have kinetic energy. Nothing can move without energy, and the faster something is moving, the more kinetic energy it has.

Potential Energy

Potential (pə ten' shəl) energy is stored-up energy. It is waiting to be in motion; then it becomes kinetic energy. A common example of potential energy is a skier perched at the top of a hill. The skier may not be moving at all, but she has *potential* energy. The moment the skier points her skis and shifts her weight in a downhill direction, gravity will pull her toward the bottom of the hill and kinetic energy will be present. Potential energy is a very useful thing.

1. Why is potential energy considered to be very useful?

   _____

2. Two types of energy are _____ and _____.

3. A definition of energy is the _____ to do _____.

4. Birds need energy to fly. Which of these is a source of energy for birds?

       wind   food   steam   instinct

5. Green plants make their own food through photosynthesis. Where do plants get the **energy** to make their own food?

       seeds   sun   water   carbon dioxide

6. An explanation that can be tested is called a(n) _____.

       hypothesis   experiment   demonstration   conclusion

7. Which of the following are *abiotic* factors that can cause change in an environment?

    flood   drought   overpopulation   photosynthesis

Remember, a solid keeps its shape and volume. A liquid keeps its volume but takes the shape of its container. A gas expands in both shape and volume to fill whatever space is available.

8. Which state of matter has a definite mass, a definite volume, but no definite shape?

    gas   liquid   solid   all three

9. The Law of Conservation of Matter says that when a sheet of paper is burned, _____.

    A) its atoms are destroyed   B) its atoms are rearranged

10. You know that burning a sheet of paper is a **chemical reaction** because during burning _____.

    A) carbon is formed and water vapor is released
    B) the paper looks different
    C) the temperature is very high
    D) the paper does not change

> The faster something is moving, the more kinetic energy it has.

Lesson #79

Here are some types of **kinetic** energy:

- **Radiant** energy moves in waves. The sun's energy is radiant. Other examples of radiant energy are light waves, radio waves, microwaves, and x-rays.

- **Thermal** energy is heat. Heat is caused by molecules moving faster and faster. The faster they move, the more kinetic energy they have and the more heat they produce.

- **Mechanical** (motion) energy is movement from something pushing against something else. Wind and waterfalls produce mechanical energy.

- **Electrical** energy travels through electrical currents. You can see electrical energy in lightning flashes during thunderstorms.

Here are some types of **potential** energy:

- **Chemical** potential energy is stored in natural resources like wood, petroleum, natural gas, and propane. The chemical potential energy is converted to kinetic energy by burning these types of fuel.

- **Gravitational** potential energy is in anything that can be dropped or can fall to Earth. Think of a roller coaster car at the peak of a hill. Even if the car is not moving at all, it has the *possibility* to move once it is tipped downward on the tracks. This is because gravity will move the car downhill. The higher up something is, the more gravitational potential energy it has.

- **Elastic** potential energy is the energy in a stretched out rubber band. If the band is stretched, you know it will snap back once it is let go. The snapping back is kinetic energy. Springs also have elastic potential energy.

Label each example as **chemical, gravitational,** or **elastic**.

1. a tank of propane on a gas grill   _____

2. a stretched out spring or rubber band   _____

3. a skateboarder at the top of a hill   _____

4. One of these terms is not like the others. Cross out the one that doesn't belong.

   streak   luster   value   hardness

5. Explain why the one you crossed out doesn't belong with the other three.

   _____

   _____

Label each example as **electrical, mechanical, radiant**, or **thermal**.

6. the rays of the sun   _____

7. heat coming from the Earth's core   _____

8. an acrobat flying through the air   _____

9. a lightning flash during a storm   _____

10. Study the diagram of a roller coaster below. At which location would a roller-car have the greatest amount of gravitational **potential energy**?

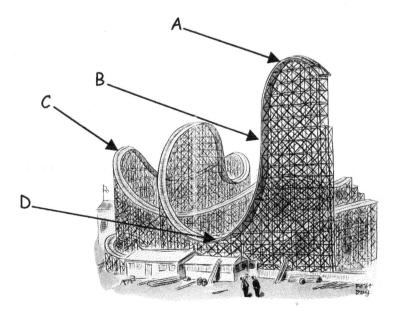

Simple Solutions© Science   Level 5

Lesson #80

1. Why do bubbles float on water?

   A) Bubbles are made of synthetic material.
   B) Bubbles are made of an organic material.
   C) Bubbles are filled with air which is less dense than water.
   D) Gravity causes bubbles to rise to the top of any liquid.

2. Color, density, melting point, and freezing point are all _____.

   A) properties of matter   C) physical changes
   B) chemical changes   D) parts of the rock cycle

3. Which of these describes a **chemical change**?

   A) ice cube melting
   B) paper shredding
   C) match burning
   D) spreading peanut butter on bread

4. According to the Law of Conservation of Matter, when paper burns, _____.

   A) some of the paper molecules disappear
   B) mass is created
   C) the molecules change but are not destroyed
   D) water is added to the environment

5. _____ is the ability to do work.

6. Which of these begins its life in water and later develops lungs and lives on land?

   humans   birds   amphibians   fish

7. Anything that can be dropped or can fall toward the surface of Earth has _____.

   A) gravitational potential energy
   B) chemical potential energy
   C) kinetic energy
   D) thermal energy

Use the diagram of an aquatic food web to answer the questions below.

8. Which organisms get energy from algae?

    crayfish   perch   sea grass   fish eggs

9. Which flow of energy is accurate?

A) sea grass – mosquito larva – fish eggs – perch
B) algae – mosquito larva – perch – fish eggs
C) algae – mosquito larva – crayfish – perch
D) mosquito larva – fish eggs – crayfish – perch

10. What may happen if the population of crayfish is wiped out?

A) The population of perch may increase.
B) The population of fish eggs may decrease.
C) The population of mosquitoes may increase.
D) The population of algae may increase.

Aquatic Food Web

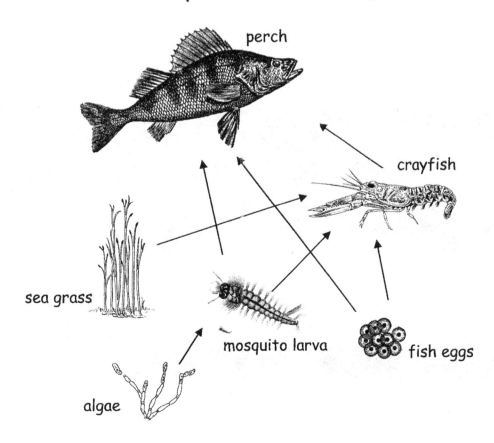

Lesson #81

Transfer of Energy

Energy is never created or destroyed, but it does change in *form*. You are already familiar with one example of this *transfer of energy*: the food chain. For example, energy from the sun goes into growing an ear of corn. You get energy from food, so eating the corn will give you energy that will enable you to do things – like reading this passage or riding your bike. That is one way energy from the sun is changed into the energy you use to do what you do.

Here are some other examples: Radiant energy from the sun can be *converted* or changed to electrical energy by using solar panels. The electrical energy can then be used to run appliances in a home or to power speakers at a music concert. Gasoline is a form of stored-up energy. When gas is put into a car's engine, the engine burns the gas as fuel. As the car moves, mechanical energy and heat energy are released. These are all examples of the **transfer of energy**.

1. Energy is neither _____ or _____.

2. Which form of energy is the result of something pushing against something else?

    light   heat   mechanical   sound

Energy comes from the sun. This energy – called solar power – can be collected using solar panels. The sun's energy can then be converted to heat or electricity.

3. What are the three states or phases of matter?

   _____   _____   _____

Write each word next to the hint that describes it.

   primary consumer   producer   decomposer
   secondary consumer   omnivore   tertiary consumer

4. breaks down organic matter   _____

5. eats animals that eat plants   _____

6. eats both plants and animals   _____

7. beginning of the food chain   _____

8. eats plants only   _____

9. eats animals that eat other animals _____

10. A combination of iron filings and sand is a mixture, not a solution, because _____.

    A) sand is too coarse to completely dissolve
    B) iron filings can be pulled out with a magnet
    C) the particles of sand and iron are not distributed evenly
    D) both sand particles and iron filings are solids

Mixtures are all around us. Some examples of mixtures are sand, soil, water, air, concrete, and wood. The particles in a mixture may not be evenly distributed.

Lesson #82

1. Which of these is an example of a **physical** change only?

    rusting iron   burning coal   melting ice   all of these

2. Which activity would produce a **chemical** change?

    painting a house   freezing berries   burning wood   all of these

3. According to the Law of Conservation of Matter, mass is neither

    _____, nor _____.

4. A **solution** is a mixture in which all of the particles are _____.

    the same   mixed evenly   easy to see   impossible to separate

5. Hummingbirds move their wings as many as 40 times per second. What kind of energy do the hummingbird's flapping wings have?

    potential   kinetic   wind   wing

Simple Solutions© Science   Level 5

6. Label each example of potential energy as **chemical**, **gravitational**, or **elastic**.

   A) a cyclist at the top of a hill   _____

   B) a battery   _____

   C) a stretched rubber band   _____

7. Label each example of kinetic energy as **radiant**, **thermal**, **mechanical**, or **electrical**.

   A) a lightning flash during a storm   _____

   B) heat generated by burning fuel   _____

   C) a bird soaring through the air   _____

   D) the sun's rays   _____

8. _____ is a measure of how closely molecules are packed in a given amount of space.

        Pressure   Density   Matter   Temperature

9. What is the best kind of soil for growing crops?

        sand   silt   soil   clay   loam

10. What is one of the characteristics of Earth as a planet that makes it possible for organisms to survive and grow?

         volcanoes   liquid water   craters   the moon

Lesson #83

Geothermal energy is heat energy. It comes from deep within the earth. (Remember, *geo* means "earth," and *thermal* means "heat.") The part of Earth that we walk around on is called the **crust**; that's the top layer. Beneath the crust is the **mantle**, a dense, mostly solid layer of rock. The center of Earth has a liquid **outer core** made of molten (melted) rock and metal called **magma** and a solid **inner core** made of iron and nickel. The inner core is the hottest part of the earth; it is almost as hot as the surface of the sun! Moving toward the center, each layer of Earth is hotter than the one above it, and the deepest layer – the core – is the hottest.

Geothermal energy captures the Earth's heat and puts it to work. This form of energy can be used to heat buildings and to produce electricity.

1. A) Which is the hottest layer of Earth?

   crust
   mantle
   outer core
   inner core

   B) Which layer is solid?

   outer core   inner core

   C) Geothermal energy can be used to _____.

   ____ heat homes

   ____ generate electricity

   ____ both of these

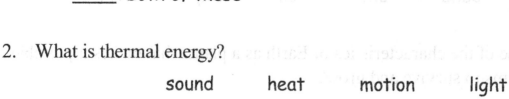

2. What is thermal energy?

   sound   heat   motion   light

3. What is magma?

   A) an erupting volcano   C) molten rock
   B) the Earth's crust   D) a type of electricity

4. The particles in a (mixture / solution) are dissolved evenly.

Simple Solutions® Science   Level 5

5. Give three examples of mixtures.

   _____   _____   _____

6. Why is clay a poor type of soil for growing plants?

   A) Clay is meant for sculpting.
   B) Clay does not allow air and water to pass through it.
   C) Clay does not hold nutrients.
   D) Clay has no decomposers in it.

7. The **transfer of energy** among organisms is called a _____.

   circuit   food chain   geothermal geyser   food source

8. A solid can change directly to a gas; this process is called _____.

   sublimation   condensation   vaporization   evaporation

9. Which of the following may decrease an ecosystem's carrying capacity? Choose all that apply.

   drought   conflict within a family   pollution

10. Which two simple machines make the pizza-cutter work?

    A) wheel-and-axle & wedge
    B) pulley & wedge
    C) screw & inclined plane
    D) ramp & pulley

Geothermal energy sometimes travels close to the surface of the earth, or it may even break through the earth's surface in an active volcano.

167

Lesson #84

Heat and Temperature

When you feel heat, you are feeling a **transfer of energy**. **Heat energy**, also called thermal energy, will move toward matter that is cooler. You already know that everything – even what you cannot see – is made of atoms, and the atoms are always vibrating. These atoms have kinetic energy, and the faster they are moving, the more kinetic energy they have. On a cold day, you may be told to shut the door. Why? Because heat will always move toward what is cooler. Thermal energy (heat) from your home will escape moving toward the cold air outside. Air, objects, and any other mass all have thermal energy. Whenever there is a difference in temperatures, thermal energy will move from hot to cold. A draft is created when cool air moves in to take the place of warmer air. But thermal energy will always move from hot to cold.

Temperature is a measure of average kinetic energy. The more kinetic energy something has, the higher its temperature will be. Temperature is measured with a thermometer. Three common units of measure are Celsius, Fahrenheit, and Kelvin. **Fahrenheit** is still the most widely used in the United States and meteorologists usually report air temperatures in both Fahrenheit and **Celsius**. Most scientists prefer the **Kelvin** scale which is based on **absolute zero** – the coldest point possible. At absolute zero, all motion stops – even the vibration of atoms. Nothing could ever be colder than that!

Scale	Freezing Point of Water	Boiling Point of Water
Celsius	0	100
Fahrenheit	32	212
Kelvin	273.15	373.15

1. What is another name for heat energy? _____

2. What is the instrument that is used to measure temperature?

   _____

Simple Solutions© Science　　　　　　　　　　　　　　　　　　　　　　Level 5

3. Heat energy will always move from _____ to _____.

4. What happens at absolute zero?

   _____

5. Which of these has the most kinetic energy?
   A) boiling water
   B) ice water
   C) water at room temperature
   D) All of these have the same amount of kinetic energy.

6. Three of these organisms have something in common. Cross out the one that doesn't belong.

   　　　spider　　　elephant　　　snake　　　rabbit

7. Explain why the one you crossed out doesn't belong with the other three.

   _____

8. Choose the correct word: (Potential / Kinetic) is stored up energy.

9. Which of these is not a solution?
   A) salt dissolved in water
   B) plant roots in water
   C) carbonated water
   D) sugar dissolved in water

> A solution is a mixture in which all of the particles are evenly distributed. A sample of the solution taken from the bottom, middle, or top of the container will be exactly the same.

10. Identify each example as either a physical or chemical change. Write **P** for physical change or **C** for chemical change.

    ____ crushing aluminum cans　　　____ decaying animals
    ____ burning trees　　　　　　　　____ recycling paper
    ____ iron rusting　　　　　　　　 ____ painting a wall

Lesson #85

1. Write each word next to the hint that describes it.

   primary consumer   producer   secondary consumer
   tertiary consumer   decomposer   omnivore

   A) breaks down organic matter _____

   B) eats animals that eat plants _____

   C) eats both plants and animals _____

   D) beginning of the food chain _____

   E) eats plants only _____

   F) eats animals that eat other animals _____

2. When water changes to ice (solid), the process is called _____.

   sublimation   condensation   freezing   melting

3. About how long does it take for the moon to go through its phases?

   21 days   365 days   30 days   7 days   24 hours

4. Which **simple machine** is pictured in the symbol?

   A) pulley
   B) motor
   C) inclined plane
   D) screw

5. When Frank's teacher lit a match, the class observed a flame and smelled sulfur. The teacher felt heat. Frank's teacher was demonstrating what kind of change?

   physical only   chemical   both   neither

6. Which of these is an example of **instinct**?

    migration   cell division   precipitation

7. The faster something is moving, the more _____ it has.

    potential energy   kinetic energy   molecules   atoms

8. Which kind of energy can be stored and used when it is needed?

    potential   kinetic   neither   both

9. One way that plant and animal cells **differ** is that _____.
    A) plant cells have chloroplasts and animal cells don't
    B) an animal cell has a nucleus
    C) plant cells contain cytoplasm
    D) animal cells have a cell membrane

10. Three rooms are next to each other with doorways in between. The temperature in each room is shown below. In which direction will thermal energy move once the doors between the rooms are opened?

Pantry	Kitchen	Den
80°	72°	60°

    A) pantry → kitchen ← den
    B) pantry ← kitchen ← den
    C) pantry → kitchen → den

You have probably been told, "Close the refrigerator door!" It's because warmer air outside the refrigerator is moving in, and that wastes energy. The refrigerator mechanism works harder to keep the temperature cool when the door is open.

Lesson #86

Transfer of Heat Energy: Conduction

Heat is transferred in three ways: conduction, convection, and radiation. **Conduction** occurs when a heat source comes into contact with other mass that is cooler. One example is holding an ice cube in your hand. What will happen? The ice cube will begin to melt right away because your hand – which is touching the ice cube – is hotter than ice. Another example is cooking pancakes. The batter is touching the hot griddle; the heat that is transferred from the griddle cooks the batter. Conduction requires contact between matter at different temperatures.

1. The faster something is moving, the more _____ it has.

    potential energy   kinetic energy

2. The transfer of thermal energy from a heat source to another object is called _____.

If you touch a flame, you will be burned because thermal energy moves from the flame which is very hot to your finger which is less hot.

3. _____ are organisms that recycle dead things by breaking them down and returning nutrients to the ecosystem.

    Garbage trucks   Scavengers   Decomposers   None of these

4. Which organism belongs to the same kingdom as mold?

    seaweed   cabbage   yeast   housefly

5. Which of these is a measure of average kinetic energy?

    air pressure   humidity   temperature   weight

Simple Solutions© Science   Level 5

6. A _____ is a simple machine that has a bar resting on a point that does not move (fulcrum). A bottle opener is an example of this simple machine.

   A) wheel-and-axle
   B) lever
   C) ramp
   D) screw
   E) pulley

7. Most of the ozone is within the _____.

   thermosphere   mesosphere   stratosphere   troposphere

8. Vertebrates that spend part of their lives living under water and later live on land are called _____.

   mammals   birds   reptiles   amphibians

9. Give the Celsius, Fahrenheit, and Kelvin temperatures at which water boils. (See Lesson #84)

   _____ °C   _____ °F   _____ K

10. Ben used the balance in his science lab to measure the mass of two marshmallows. He recorded the mass; then he roasted the marshmallows over a campfire. When Ben put the marshmallows back on the balance, the measurement of mass was much less than it had been before roasting. What is the best explanation for what happened to the mass of the marshmallows?

    A) Some of the mass was destroyed during roasting.
    B) Burning created new matter which had less mass.
    C) Some gasses were released during the roasting.
    D) Ben used the balance incorrectly.

Lesson #87

Transfer of Heat Energy: Convection and Radiation

You already know that warm air rises, and that will help you to understand convection. When gases or liquids heat up, they become less dense and move upward. Cooler gases and liquids sink or fall below a rising warmer mass. **Convection** is the circulation of a mass of liquid or gas, rising as it is heated and sinking as it is cooled. This warmer gas or liquid heats the air or liquid above it. When water is heated on the stove, it rises to the top of the pot; the cooler water moves below. Warm air from a baseboard heater rises to heat a room – cooler air sinks and then rises after it is heated.

Thermal energy is also transferred by **radiation**. Radiant energy moves in waves. Waves from the sun carry energy. If you step on a sunny sidewalk in your bare feet, you will notice that it is hot while a nearby sidewalk in the shade is cooler. That's because the sun's rays are delivering heat as well as light. Other examples of radiant energy are light waves, radio waves, microwaves, and x-rays. **Conduction** requires contact between two objects, and **convection** requires a gas or liquid to move through. **Radiation** is the transfer of energy using waves.

Read each example of a transfer of thermal energy. Write **conduction**, **convection**, or **radiation** to identify the type of movement.

1. _____ warm air mass rising

2. _____ heat from the sun

3. _____ touching a hot stove

4. _____ a hot sidewalk on a sunny day

5. _____ pressing a shirt with a hot iron

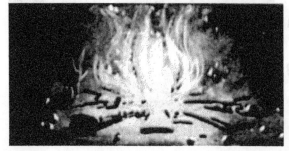

Light and heat from a campfire move in waves. This is an example of radiation.

A hot air balloon uses convection. Fire heats the air forcing it upward, and that makes the balloon rise.

6. One *distinction* between a plant cell and an animal cell is _____
   A) a plant cell has chloroplasts; an animal cell does not.
   B) a plant cell has a nucleus; an animal cell does not.
   C) only animal cells contain cytoplasm.
   D) a plant cell has a cell membrane; an animal cell doesn't.

7. A(n) _____ is a statement of a possible answer to a scientific question or problem.

   observation   variable   experiment   hypothesis

8. Why is it important to measure the speed of wind?
   A) Changes in wind speed may signal changes in weather.
   B) It is a meteorologist's job to measure wind speed.
   C) There are ways to slow the wind down.
   D) High winds may signal a volcanic eruption.

   An **anemometer** measures the speed of wind.

9. Mr. Welch sliced a loaf of fresh raisin bread. Which of the following is true?
   A) The unsliced bread had greater mass.
   B) The sliced loaf has greater mass.
   C) The bread has the same mass before and after slicing.
   D) Raisin bread does not have mass.

10. The particles that make up a gas _____.
    A) move slowly and have very little space between them
    B) are bound together and will not break apart
    C) are the same as particles in a solid or liquid
    D) move fast and have a lot of space between them

Lesson #88

1. Mindy used a thermometer to measure the temperature of three liquids. Which of the liquids had the **greatest average kinetic energy**?

   A) milk: 36°
   B) coffee: 200° F
   C) orange juice: 40° F
   D) Kinetic energy is the same in all three liquids.

2. Water will change from liquid to solid when it reaches its _____ point.

   boiling   condensation   melting   freezing

3. Why are you able to cool off by jumping into a swimming pool on a very hot day?

   A) Heat is transferred from your body to the cool water.
   B) Coolness is transferred from the water to your body.
   C) Water is always cooler than air.
   D) All of the above.

4. Scientists use a system of _____ to arrange organisms into easy-to-manage groups.

   A) observation
   B) investigation
   C) classification

5. Which is an example of a physical change?

   A) forming an object from a lump of clay
   B) burning a piece of paper
   C) digestion
   D) corrosion

On a cool day, heat from your home will move out of openings in windows and doors. On a warm day, heat will move from outside into the house. Thermal energy always travels from warmer areas to cooler areas.

Simple Solutions© Science   Level 5

6 – 10. The chart below shows some organisms in an ecosystem and what those organisms eat. In the box below, draw a food web that includes all the organisms in the chart. Use words, symbols, or pictures.

Organism	What it Eats	Picture
Owl	frog, centipede	
Frog	ant, moth, dragonfly	
Ant	sedge	
Moth	sedge	
Dragonfly	sedge	
Sedge	xxx	
Centipede	ant	

Lesson #89

Natural Resources

Everything you see around you comes from natural resources. Air, water, soil, animals, wood, cotton, metal, and rubber are just a few examples. Even manufactured items are made from natural resources. All the energy that you use every day and all the energy that is used to make things also comes from natural resources. All of these resources are either renewable or nonrenewable.

A **renewable resource** is one that can be replenished or replaced in a "short" period of time – about fifty to a hundred years. Renewable resources are replaced by *natural ecological cycles* like the water cycle, the carbon cycle, or the oxygen cycle. And when renewable resources are used wisely, they can be used over and over again without ever running out. Water is an example. If water is used carefully, and the water source is protected, the water will always be reusable. Sometimes, however, a water source becomes contaminated by industrial waste, pesticides, fertilizers, or sewage (human waste). Then the water is not reusable. Other renewable resources – like forests, animals, and even air – can also be overused or mismanaged. This makes them unavailable to reuse.

A **nonrenewable resource** cannot be replaced within fifty or even one hundred years. Once a nonrenewable resource is used up, it is gone and can never be replaced in a person's lifetime. Natural resources that we call **fossil fuels**, for example, took *hundreds of millions of years* to form. Coal, oil, and natural gas are fossil fuels. Other nonrenewable resources are minerals and old-growth forests (like the redwood forests). Soil may be considered a nonrenewable resource. A large part of soil is weathered rock, and weathering is a process that takes thousands of years. However, soil – if used carefully – can replenish its other parts (organic matter, water, and air) within a shorter period of time.

Pollution from factories may enter rivers, streams, or lakes; it seeps into groundwater, and groundwater flows toward lakes and oceans. This is one way that toxic waste can enter the water cycle.

1. Natural resources that can be replaced by natural ecological cycles are _____.

   renewable resources   nonrenewable resources   both of these

2. Which of these are **nonrenewable** natural resources? Underline all that are.

   wind   coal   water   natural gas   diamonds

3. What would make a renewable resource unusable or unavailable?

   pollution   contamination   overuse   all of these

True or False?

4. _____ One of the physical characteristics of a solid is that its atoms do not vibrate.

5. _____ The atoms in liquids and gases are free to move around.

6. A solution is called **homogeneous** because _____.
   A) its parts have been heated
   B) all solutions are the same
   C) it comes from a mixture
   D) all of its particles are evenly distributed

7. The densest layer of Earth's atmosphere is the _____.

   thermosphere   mesosphere   stratosphere   troposphere

8. Which of these is an example of **radiant** energy?

   sunlight   a kite flying   electricity   a battery

9. Which word means *heat from the Earth*?

   geology   extraterrestrial   geothermal   thermodynamic

10. Which pair of organisms is most closely related?

    rosebush/spider   jellyfish/bumblebee   mushroom/hawk

Lesson #90

Energy Comes From Natural Resources

The energy that we use in our homes, at school, in our cars, and everywhere else comes from natural resources. Much of our energy comes from **fossil fuels** (coal, oil, natural gas) which are nonrenewable. The most abundant fossil fuel is **coal** which can be burned to generate electricity. **Oil** is used to make petroleum products like gasoline, diesel fuel, and propane. Oil products are used to run cars, to fly airplanes, and to power large machinery. **Natural gas** is used to heat over half of all the homes in the United States. It is also used for cooking and running household appliances. Natural gas is used quite a bit in industry, and it is even used to make electricity.

Most Americans and many people all over the world depend upon nonrenewable resources to supply energy. However, there are some serious disadvantages to using nonrenewable resources like fossil fuels. First, these resources are limited and will definitely run out. Fossil fuels release harmful pollutants into the environment. Coal mining can be very dangerous to miners. Large areas of land are damaged during the mining of both coal and oil. Mining can also cause water pollution since mines come into contact with rivers, lakes, and streams. These disadvantages have prompted people to explore ways to conserve all natural resources and to develop ways to get energy from renewable resources.

1. Natural resources that can be replaced within fifty to one hundred years are called _____.

    renewable resources   nonrenewable resources   both of these

2. Name three fossil fuels.

    _____ _____ _____.

3. Which fossil fuel is used mostly for transportation?

    _____

During mining, land is stripped. The land may or may not be restored after the mining is done.

Simple Solutions© Science   Level 5

4. Every plant and animal cell has a _____ that directs all of the cell's activities.

5. About how long does it take for the moon to go through its phases?

    21 days   365 days   30 days   7 days   24 hours

6. Jenny made a steaming hot cup of tea, and left it on the kitchen table. An hour later, the teacup was cool to the touch. Why did the tea cool off?

   A) Thermal energy moved toward the cooler air in the kitchen.
   B) Coolness from the room moved into the teacup.
   C) Radiation caused the cup to cool off.
   D) Heat moved from the table to the cup by conduction.

7. Which of these are naturally occurring, inorganic solids?

    polymers   fungi   minerals   atmospheric gases

8. Ice will become liquid water when it reaches its _____ point.

    condensation   boiling   melting

9 – 10. Explain what scavengers are and why they are important to an ecosystem. Use some of the words from the word bank.

animals	decaying	environment	food chain
clean	nutrients	organisms	organic

_____

_____

_____

_____

_____

Lesson #91

Energy Comes From Natural Resources

Most of the energy used by people in the United States comes from nonrenewable resources. However, some renewable resources are also being used. Wind, solar, geothermal, water, and biomass are renewable natural resources. The uneven heating of Earth's surface creates **wind**, and you already know that large air masses are constantly on the move in our atmosphere. Wind can generate electricity. **Solar** energy can also generate electricity or heat. Solar panels and solar collectors get energy from the sun. **Geothermal** energy comes from the Earth. As you know, Earth's core is hotter than the surface of the sun. In addition to that, radiant heat from the sun is absorbed by Earth's crust, and decaying organic matter in the crust also generates heat. Geothermal energy is used for heat and to generate electricity. **Hydropower** or **water** power can generate electricity. The water must be moving rapidly and in very great amounts (like at Niagara Falls). **Biomass** is organic material (*bio* means "life" and *mass* is anything that takes up space and has volume). Living things absorb energy from the sun, and that stored-up energy can be released as heat when biomass is burned. Some examples of biomass are wood, manure (animal waste), and crops like corn. Even organic *garbage* can be burned to release energy.

1. What is one disadvantage to using wind to generate electricity?

   A) Wind power pollutes the environment.
   B) Wind is a nonrenewable resource.
   C) Collecting wind energy is very noisy.
   D) Sometimes the wind isn't blowing in certain places.

2. How does biomass convert to energy that people can use?

   A) Biomass decays.
   B) Energy is released when biomass is burned.
   C) Biomass puts nutrients back into the soil.
   D) Decaying plants become garbage.

Niagara Falls is a great place to harness hydroelectric power. As much as 600,000 gallons of water per second spill over the cliffs of Niagara.

Simple Solutions© Science   Level 5

3. In the United States, most of our energy comes from (renewable / nonrenewable) resources.

4. Which two simple machines allow the hand truck to move easily with a heavy load?

   wheel-and-axle   lever   wedge

   ramp   screw   pulley

5. Photosynthesis is a <u>chemical</u> reaction because _____.

   A) During photosynthesis a new substance, called glucose, is formed.
   B) Photosynthesis occurs in nature.
   C) Only green plants can perform photosynthesis.
   D) The process can never be seen or reversed.

6. While this player is holding the basketball, what kind of energy does it have?

   kinetic   potential   thermal   motion

7. Which of these is a **producer**?

   butterfly   soil   bean plant   sun

8. Which is true?

   A) Rocks contain minerals.
   B) Minerals are rocks.
   C) Both are true.

9 – 10. Explain two disadvantages of using fossil fuels. (See Lesson #90.)

_____

_____

_____

_____

Simple Solutions© Science Level 5

Lesson #92

Conserving Energy Resources

Many natural resources are not renewable. And those that are renewable can easily become unfit for use if they are not protected. **Energy conservation** and **energy efficiency** are two ways to protect and preserve natural resources. You may have heard the phrase, "Reduce, Reuse, and Recycle." The three R's are a way of conserving energy by conserving natural resources. Another word for *conserving* is *saving*.

You can save energy by turning off lights that are not in use, turning off the water faucet while brushing your teeth, or riding your bike (instead of going in a car). These are ways to **reduce** the amount of energy resources you use. You can save energy by writing on both sides of a sheet of paper before getting a new one, donating used items like clothing or toys instead of throwing them away, or using regular reusable silverware instead of disposable plastic. These are ways to **reuse** items. You can also **recycle** to save energy. These are materials that may be recycled in your community: plastic, aluminum, cardboard, newspapers, and magazines. Once you know which items can be recycled in your community, always collect those items in recycling bins.

Energy efficiency is also a way to **reduce** the amount of energy resources you use. When something is *efficient*, it is not wasteful. A household, school, or business can be more efficient by using compact fluorescent light bulbs, water-saving toilets, and energy efficient appliances. Other ways to be energy efficient include carpooling, insulating and weatherproofing doors and windows, and using high efficiency appliances.

1. List three materials that can be recycled.

   _____   _____   _____

2. Why is it important to protect and preserve natural resources?

   A) People rely on natural resources for energy.
   B) Natural resources like lakes and forests provide recreation.
   C) Everything in the environment is linked together.
   D) All of these are reasons to conserve natural resources.

3. Three of these terms have something in common. Cross out the one that doesn't belong.

         wind   coal   natural gas   oil

Simple Solutions© Science   Level 5

4. Explain why the one you crossed out in item 3 doesn't belong with the others.

   _____

5. Which organism belongs to the same kingdom as mold?

   apple tree   butterfly   cactus   mildew

6. When a liquid changes to a gas, the process is called _____.

   sublimation   evaporation   melting   watering

7. Earth's atmosphere has the right mix of gases needed to support life as we know it. What are three of these gases? (See Lesson #49.)

   A) nitrogen, oxygen, and carbon dioxide
   B) carbon monoxide, nitrogen, and ozone
   C) oxygen, sulfur dioxide, and methane
   D) nitrogen, methane, and carbon monoxide

8. (Potential / Kinetic) energy is the energy of motion.

9. A warmer air mass will (rise / sink), and a cooler air mass will (rise / sink).

10. When Julie dropped a seltzer tablet into a glass of water, she observed a chemical reaction. The mixture of the water and the tablet released carbon dioxide (she saw bubbles). Which of the following is true?

    A) Matter was neither created nor destroyed in this event.
    B) The chemical reaction caused a change in the water and seltzer tablet.
    C) The carbon dioxide and water will have the same mass as the water and seltzer tablet had.
    D) All of these statements are true.

Lesson #93

Dr. Maria Telkes, Physical Chemist & Inventor (1900 – 1995)

Maria Telkes was born in 1900, long before **environmentalists** began to realize that Earth's supply of fossil fuels would one day run out. As a young girl growing up in Budapest, Hungary, Maria felt the warmth of the sun and was spellbound by its light, heat, and movement. She imagined a machine that could capture the energy of the sun. By the time she was in high school, Maria began to read everything she could find about the sun. When she finished reading books in Hungarian, she looked for books that were written in other languages. When she finished high school, Maria went to college. After many years of study, she earned a doctorate in **physical chemistry**. Then Maria moved to the United States, and she became an American citizen in 1937.

As a newcomer, Maria worked at the Cleveland Clinic Foundation. She was a **biophysicist**. Later, she worked at Westinghouse Electric as a **research engineer**. **Physics** is the study of matter and energy; **biology** is the study of living things. As a biophysicist, Maria studied how energy affects living organisms. As a research engineer, she worked on transforming heat energy into electrical energy. With all of this education and experience, Maria later moved into the career that made her famous. While working at the Massachusetts Institute of Technology (MIT), she developed a design for a solar heated house. Her design was successful, and she went on to develop other mechanisms like an easy-to-use solar oven and a water distiller. A water distiller made it possible to transform saltwater into drinkable water. These inventions are still used today. Maria Telkes received many honors for her outstanding achievements and has earned the nickname, "the Sun Queen."

1. How is the work of Maria Telkes related to **conservation**?

    A) Maria's work makes it possible to use renewable resources.
    B) Maria came to the United States from Budapest, Hungary.
    C) Maria was able to read books in many languages.
    D) all of the above

2. What did Maria Telkes do as a research engineer?

    _____

3. Which career combines the study of living things with the study of energy and matter?

   biophysicist   engineer   environmentalist   conservationist

4. Why is a water distillation mechanism an important invention?
   A) It is worth a lot of money.
   B) It converts seawater to fresh water.
   C) It collects heat from the sun.
   D) none of these

5. One example of how organisms interact with one another in an environment is a(n) _____.

   volcanic eruption   weather event   food web   earthquake

6. Below is a list of fossil fuels. Cross out any that don't belong.

   coal   oil   wind   natural gas   water

7. Which two things are needed for every experiment?

   materials list   goggles   procedures   lab sheet   timer

8. What is biomass?
   A) a fossil fuel
   B) a nonrenewable resource
   C) a type of energy extracted from coal
   D) wood, manure, crops, and organic garbage

> Soil is a natural resource. It contains weathered rock, and weathering is a process that takes thousands of years. However, if soil is used wisely, its other parts (organic matter, water, and air) can be replenished.

9 – 10. Is soil a renewable or a nonrenewable resource? Write your explanation below. (See Lesson #89.)

_____

_____

_____

Lesson #94

Atoms and Molecules

All matter is made of tiny parts called atoms. Even cells are made of atoms, so both living and non-living things are made of atoms. An atom is so small that it cannot even be seen with a microscope, and millions of atoms could fit on the head of a pin. Atoms are the building blocks of **elements**. Each element is made up of only one kind of atom. For example, oxygen is made of only oxygen atoms; hydrogen is made up entirely of hydrogen atoms, and silver is made of silver atoms. All the elements that we know of are listed in the Periodic Table of Elements. An **atom** is the smallest bit of any given type of matter.

When the atoms of two or more different elements join together, that makes a **molecule**. For example, both hydrogen and oxygen are elements, but when two hydrogen atoms join with an oxygen atom, that makes a water molecule (H_2O). A water molecule is not the same as hydrogen or oxygen; it has different properties. You normally experience hydrogen and oxygen as gasses, yet water is a liquid under the same conditions. Another example is table sugar. Sugar molecules are made of carbon, hydrogen, and oxygen. You may know that carbon is the black substance you write with when you use a pencil. Charcoal is also mostly carbon. At room temperature, oxygen and hydrogen are both gases. But these three elements can go together in this formula: $C_{12}H_{22}O_{11}$, and that makes a sugar molecule. Table sugar is a lot of white crystals that definitely do not have the same properties as carbon, hydrogen, and oxygen.

1. Hydrogen is one of the building blocks of a water molecule. Which of these is true?

    A) Hydrogen and water have different properties.
    B) Water is an element; hydrogen is an atom.
    C) Hydrogen is made of atoms; water is not.
    D) Water molecules and hydrogen atoms are the same.

2. All the elements that we know of are listed where?

Carbon atoms are in lumps of charcoal and in lumps of sugar.

Simple Solutions© Science   Level 5

Fill in the blanks using terms from the word bank. Some of the words will not be used.

  atoms   carbon   cells   elements
  hydrogen   molecule   oxygen   properties

3. All matter is made of _____.

4. A _____ is the combination of two or more different types of atoms.

5. _____ are made up of only one type of atom.

6. Which of these are **renewable** natural resources? Underline all that are.

  fish   coal   trees   petroleum   oxygen

7. A series of steps used to investigate a scientific question, phenomenon, or problem is _____.

  the scientific method   a hypothesis   a conclusion   a variable

8. Behaviors that animals have without being taught are _____.

  metamorphosis   instincts   training   endangered

9. Soil is a(n) _____.

  element   mineral   solution   mixture

10. The troposphere contains _____ of the gases in the atmosphere.

  25%   90%   10%   100%

The Periodic Table of Elements is an organized list of all the known elements.

189

Lesson #95

Parts of an Atom

Even though atoms are too small to see, scientists know that an atom is made of even smaller particles called protons, neutrons, and electrons. Each atom has a nucleus which is made of protons and neutrons. Electrons spin around the nucleus in an orbit (like a bee hovering near a flower). The electrons are held in this orbit by an electrical force.

Protons and electrons carry an electrical charge: **Protons** have a *positive* electrical charge; **electrons** have a *negative* electrical charge. And **neutrons** are *neutral* (have neither a positive nor a negative charge). When an atom has an equal number of protons and electrons, the positives balance out the negatives — they are neutral.

Fill in the blanks. Write **negative, positive,** or **neutral**.

1. **Protons** have a _____ electrical charge.

   Electrons have a _____ electrical charge.

   Neutrons have a _____ electrical charge.

2. Many atoms are neutral. This means they have _____.

   A) an equal number of protons and electrons
   B) mostly neutrons
   C) more electrons than protons
   D) more protons than electrons

3. Which of these is a solution?

   salad dressing   saltwater   pond water   cereal

4. An herbivore will never eat _____.

   plants   animals   seeds   nuts

   > The nucleus is a dense area of protons and neutrons. Electrons spin in a wide orbit around the nucleus. An atom is mostly empty space.

190

Simple Solutions© Science   Level 5

5. _____ is the weight of the atmosphere pressing toward Earth's core.

   Air pressure   Ozone   Prevailing winds   Gravity

6. Three of these terms name types of precipitation. Cross out the one that doesn't belong.

   rain   dew   snow   hail

7. Which pair of organisms is most closely related?

   cockroach/salamander   worm/mold   fish/amoeba

8. This rock formation was most likely created by which of these?

   A) evaporation
   B) moving water
   C) plant growth
   D) condensation

9. What are the three states of matter?

   _____   _____   _____

10. Why are scientists working to develop alternative energy sources like wind, water, solar, geothermal, and biomass?

    A) These energy resources are renewable.
    B) Earth's supply of coal, oil, and natural gas will eventually run out.
    C) Burning fossil fuels causes pollution.
    D) All of the above are reasons to develop alternative energy sources.

Wind machines collect the wind's kinetic energy and enable it to be converted to electrical energy.

Lesson #96

1. A _____ is two or more atoms joined together.

2. An atom is considered electrically balanced or neutral when _____.
   A) the number of protons is greater than the number of electrons
   B) the number of electrons is greater than the number of protons
   C) there are no neutrons present in the atom
   D) the number of protons is equal to the number of electrons

3. This geyser is releasing geothermal energy above ground. In which direction will the geothermal energy move through cooler air?

   upward   downward   sideways   no movement

4. The troposphere is the densest layer of the atmosphere. Why?
   A) It is the highest layer of the atmosphere.
   B) The upper layers have more oxygen than the troposphere.
   C) The weight of the upper layers presses down on the troposphere.
   D) The troposphere has fewer gas particles than the upper layers.

5. Three of these terms have something in common. Cross out the one that doesn't belong.

   microscope   thermometer   barometer   hygrometer

6. Explain why the one you crossed out doesn't belong with the other three.

   _____

   _____

7. All of the balls on these shelves have **gravitational potential energy**. Which shelf holds the balls with the <u>most</u> potential energy?

   A) basketballs
   B) assorted balls
   C) footballs
   D) volleyballs

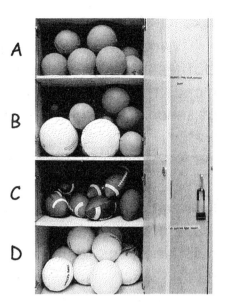

8 – 10. Complete the chart below by filling in a drawing and a description of each type of cloud.

Cloud Name	Illustration	Description
cirrus		
cumulus		
stratus		

Lesson #97

Electricity

Watching television, playing a video game, surfing the internet, tumbling clothes in a dryer – what do these activities have in common? Electricity makes them possible. We depend upon electricity, and we use it at home, at school, in businesses, and in industry every day. Electrical energy heats water and cools air; it runs machinery and appliances, and it lights up buildings, runways, and streets. What is electricity, and where does it come from?

Electricity is a form of energy produced by the movement of electrons. Remember, all matter is made of tiny particles called atoms. Atoms are made of even smaller particles: protons and neutrons which form the nucleus, plus electrons which orbit around the nucleus of an atom. Protons have a positive charge, and electrons have a negative charge (neutrons have a neutral charge). Most atoms have an equal number of protons and electrons, making them electrically balanced or *neutral*. Sometimes, however, electrons free themselves, moving out of their orbit and away from the atom. If an atom has more protons than electrons, it has a *positive* electrical charge. If an atom has more electrons than protons, it has a *negative* electrical charge.

1. Electricity is produced by the movement of _____.

2. An atom that is **neutral** has an equal number of _____.

    protons and neutrons   protons and electrons   neutrons and electrons

3. An atom will have a **negative** charge when _____.
    A) it has the same number of protons, neutrons, and electrons
    B) it has no neutrons
    C) it has more protons than electrons
    D) it has more electrons than protons

4. Give three examples of how you use electricity every day.

    _____

    _____

    _____

5. What are the four basic needs of animals?

   _____   _____

   _____   _____

6. The photo shows that a _____ keeps its definite shape and volume whether it is in a container or not.

   A) gas
   B) liquid
   C) solid
   D) all three

7. Which of these are processes that shape the surface of the Earth?

   condensation   shifting of tectonic plates   cell division   erosion

8. The word *carnivore* means _____.

   meat-eater   plant-eater   producer   animal

9. Natural resources that can be used over and over again, if used wisely, are called _____.

   renewable resources   nonrenewable resources   both of these

10. Which simple machine is made with a rope, chain, or belt wrapped around a curved wheel?

    screw   wedge   inclined plane
    pulley   ramp   wheel-and-axle

Lesson #98

Static Electricity

You experience static electricity as charged particles moving between objects. For example, you can make a balloon stick to the wall by rubbing it against a woolen blanket or sweater. Why does this work? The balloon is made of atoms, and its atoms have protons, neutrons, and electrons. Rubbing the balloon against fabric frees some electrons. The balloon picks up these extra electrons and becomes negatively charged – it gains static electricity. The balloon then sticks to the wall because the negatively charged atoms that make up the balloon are attracted to the positive charges in the wall. Walking on carpet in your socks can generate static electricity in the same way. Your body picks up extra electrons that are set free by the friction of socks against carpet. When you touch a light switch or even another person, electrons jump from you to whatever you touch. It can give you a little shock, and you may even see a spark.

Lightning during a thunderstorm is another example of static electricity. Ice crystals and water droplets rub against each other, creating a negative charge near the bottom of a cloud. A positive charge collects at the top of the cloud. Electrons jump from one cloud to another or from a cloud to an object on the surface of the Earth. This creates a huge spark of lightning.

Static means "still," and **static electricity** is electricity at rest. Static electricity is a type of *potential energy*. The energy becomes *kinetic* when the static charge is released – as in a slight shock, a spark, or a lightning flash.

1. Look at the diagram below. What will most likely happen once the atoms in the balloon become *neutral*?

   A) Nothing will happen.
   B) The balloon will give off sparks.
   C) The balloon will drop to the floor.
   D) The balloon will burst.

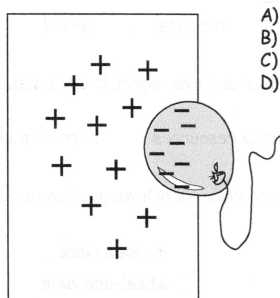

Negatively charged atoms in the balloon are attracted to positive charges in the wall. So the balloon sticks.

Simple Solutions© Science   Level 5

2. **Static electricity** is created by _____
   A) objects exchanging electrons.   C) batteries.
   B) the electric company.   D) none of these.

3. What should you do during a thunderstorm or electrical storm?
   nothing   seek shelter   talk on the telephone   take a swim

4. _____ is anything that takes up space and has mass.

The following are examples of physical and chemical changes:
   A) mixing yeast and sugar   C) slicing bread   E) baking bread
   B) dissolving sugar in water   D) freezing water   F) melting cheese

5. Write the letter of each action that creates a **chemical** change.

   _____

6. Write the letter of each action that creates a **physical** change.

   _____

7. Another word for **thermal** energy is _____ energy.

8 – 10. Explain how the decay of this tree is causing change in its ecosystem.

197

Simple Solutions© Science — Level 5

Lesson #99

Current Electricity

You learned that electricity is produced by the movement of electrons. Static electricity is a type of potential energy. When it is released as kinetic energy, static electricity creates only a short charge or burst of energy. This is not enough energy to run appliances, light schoolrooms, or power a manufacturing plant. Another type of electricity is **current electricity**. The electrons in current electricity flow in a constant stream of energy. Current electricity is *kinetic*. It flows along metal wires which are excellent conductors. An electrical **conductor** is a material that allows electricity to move easily through it. Most metals are very good conductors. In order for electricity to be useful to consumers, it must be delivered in a constant, steady flow.

Current electricity can be generated at electric power plants and can then be delivered to homes, offices, schools, and other places through copper wire. Consumers make use of current electricity by plugging cords into power outlets. When a cord is plugged into an electrical outlet, the flow of electricity continues through the cord to operate the electrical appliance. Current electricity can be converted into other types of energy such as light, heat, and sound.

1. Which of these is not safe to do and may cause electric shock?

    A) climbing playground equipment
    B) playing basketball outdoors
    C) climbing utility poles
    D) attaching a swing to a tree branch

2. Electricity at rest is _____ electricity.

    neutral   charged   static   current

3. What is one distinction between static electricity and current electricity?

    A) Both are exactly the same.
    B) Static electricity is potential; current electricity is kinetic.
    C) Static electricity is kinetic; current electricity is potential.
    D) Static electricity is dangerous; current electricity is not.

4. A cow, a chicken, and a horse all belong to the same **kingdom** because they _____.

    A) all live on farms
    B) can make their own food
    C) are multi-celled consumers
    D) can be seen without a microscope

5. To investigate a problem, question, or phenomenon, scientists use the _____.

6. The Law of Conservation of Matter states that mass is neither created nor destroyed. If the mass of a substance measures less after a chemical change, what is the most likely explanation?
   A) Some of the mass was destroyed during the chemical reaction.
   B) There is less matter as a result of the chemical change.
   C) Some gases were released during the chemical reaction, and these were not included in the measurements.
   D) Mass cannot really be measured.

7. The densest layer of Earth's atmosphere is the _____.

   thermosphere   mesosphere   stratosphere   troposphere

8. Which of these is a renewable energy resource?

   waterpower   oil   natural gas   coal

9. Look at the two socks in the diagram below. If they were just taken out of the dryer, what kind of electricity do they hold?

   static   current   neither   both

10. One sock has a positive charge and the other has a negative charge. What will happen if these two socks get close to each other?
    A) Nothing will happen.
    B) The socks will cling to each other.
    C) The socks will fade.
    D) The socks will repel each other.

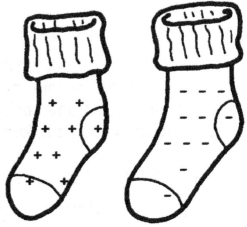

Lesson #100

Conductors and Insulators

You learned that an electrical **conductor** is a material that allows electricity to move easily through it. Metals, like copper, are excellent conductors of electricity. Not all conductors allow electricity to flow equally, but metals are the best. Other conductors are live trees (because of the sap inside them), the water in swimming pools, and the human body. Electrical currents will pass through all of these conductors, and that is why you should go indoors when you see lightning. You should never go near a power line that is sagging or touching the ground, and of course, never use any electrical appliance near a sink or while taking a bath.

Any material that is a poor conductor of electricity is called an **insulator** (or a non-conductor). Plastic, rubber, dried wood, and glass are good insulators; these materials can protect consumers against electric shock and injury. The electrical cord that you plug into an outlet is rubberized or plastic on the outside; the metal wire is inside. Loose electrical currents can hurt people and cause fires. Cords that are cracked or fraying should never be used to carry electricity.

Conductors: aluminum, brass, bronze, concrete, copper, gold, iron, mercury, silver, steel, graphite, water-based solutions

Insulators / Non-conductors: asphalt, ceramic, dried wood, fiberglass, glass, oil, paper, plastic, porcelain, rubber, pure water

1. Which of these is not a safe practice during a thunder and lightning storm?

    swimming outdoors   watching TV   playing cards   doing homework

2. Which of these is a good conductor of electricity?

    distilled water   water in a puddle

3. Why is "live wood" (as in a tree) a good conductor while dried wood is not?

    A) There are no electrons in dead wood.
    B) Live trees contain sap.
    C) Live trees stand upright.
    D) Lightning only strikes live trees.

4. The electrons in _____ electricity flow in a constant stream of energy.

    static   current   neither   both

Simple Solutions© Science Level 5

5. Which of the following are *abiotic* factors that can cause change in an environment?

   plant overgrowth   erosion   birds nesting   hurricane

6. Which of these is a solution?

   vinegar and oil   sand and silt   food coloring and water

7. The photos show that neither a _____ nor a _____ have a definite shape.

8. When an ice cube melts, there is a physical change, not a chemical change. You know this because _____.

   A) ice and liquid water are the same molecules
   B) water always freezes at 32°F
   C) water is made of elements
   D) water is never part of a chemical change

9. Three of these terms have something in common. Which one doesn't belong?

   screw   wedge   pulley   motor

10. Explain why the one you chose in item 9 doesn't belong with the other three.

_____

_____

201

Lesson #101

Simple Circuits

As you know, current electricity flows constantly along a path. The path that the electricity follows is called a **circuit**. Every circuit has a **source** (which supplies electricity) and a **conductor** (pathway) for electricity to travel along. The source may be a battery, a generator, or an electrical outlet. The conductor is usually an insulated wire. Most circuits also contain an **appliance**, which refers to any device that is powered by electricity: a light bulb, a hair dryer, a TV, or a computer, for example. One other useful device that may be part of a circuit is a switch. A **switch** opens or closes the circuit. When a power switch is ON, the circuit is closed or **complete**. This allows the electricity to flow freely through the circuit. When the switch is OFF, the circuit is open or incomplete and will not allow electricity to flow.

You can change a circuit by adding or taking away parts. Look at the illustration below. You can make the conductor longer or shorter by adding more wire or making the wire shorter. If you add more wire, the bulb will dim because the current has to travel farther. If you make the wire shorter, the bulb will be brighter. You can add more bulbs. If you keep the source the same (just one battery) and add light bulbs, the bulbs will all be dimmer. This is because more bulbs are sharing the same amount of current. Also, you can attach additional batteries (increase the source). This will make the bulb brighter since you are adding more electrical current. If you add too much current, the bulb may burn out.

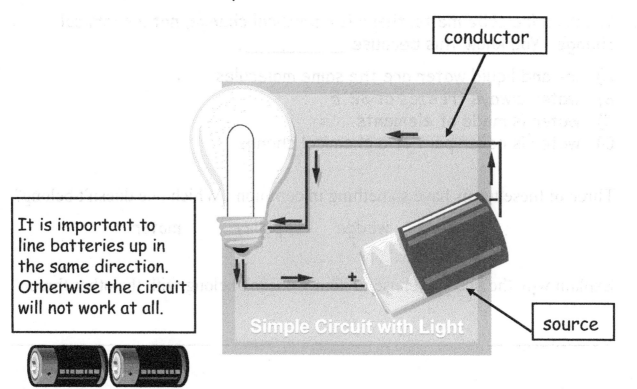

It is important to line batteries up in the same direction. Otherwise the circuit will not work at all.

Simple Circuit with Light

Simple Solutions© Science   Level 5

1. What are two parts of every simple circuit?

   _____   _____

2. List four items in your home or school that are powered by electricity.

   _____   _____

   _____   _____

Study the drawing of a simple circuit and answer the next three questions.

3. If more bulbs are added to the circuit, what will happen?

   A) The bulbs will get brighter.   C) The circuit will no longer work.
   B) The bulbs will get dimmer.

4. If more batteries are added to the circuit, what will happen?

   A) The bulbs will get brighter.   C) Nothing will happen.
   B) The bulbs will get dimmer.

5. If more wire is added to the circuit, what will happen?

   A) The bulbs will burn out.   C) The bulbs will get dimmer.
   B) The bulbs will get brighter.

6. A (gas / solid) retains neither its shape nor its volume outside of a container.

Write the word next to the hint that describes it. (Not all words will be used.)

   tertiary consumer   herbivore   secondary consumer
   producer   decomposer   omnivore

7. breaks down organic matter   _____

8. eats both plants and animals   _____

9. makes its own food   _____

10. eats plants only   _____

Lesson #102

Circuits and Switches

Only a **closed circuit** will work properly. A switch opens or closes a circuit. When a switch is turned ON, the circuit is closed. When a switch is turned OFF, the circuit is open or broken. An open or broken circuit will not work because the current cannot flow through. Just about everything that uses electricity uses switches. The keys on computer keyboards or calculators and the buttons on video games are all types of switches.

1. When a circuit is **closed**, _____.

    A) power will flow through the circuit
    B) the circuit is open or broken
    C) the circuit will not work properly
    D) the circuit may or may not work

2. When a switch is turned OFF, _____.

    A) the circuit is closed and will not work
    B) the circuit is open and will not work
    C) power will flow freely through the circuit
    D) there is a complete circuit

Electrical switches can be pull chains or wall buttons.

3. Which of these **does not** use a switch?

    battery   computer keyboard   TV   cell phone   lamp

4. Sometimes loose electrons move through matter. This creates _____.

    electrical currents   new atoms   new elements

The keys on computer keyboards or calculators and the buttons on video games are all types of switches.

Simple Solutions© Science Level 5

5. Which pair of organisms is most closely related?

    octopus/earthworm   salamander/seaweed   sunflower/lobster

6. Oxygen, hydrogen, and carbon are all examples of _____.

    cells   elements   molecules   gases

7. Which of these is a solution of gas dissolved in a liquid?

    carbonated water   salad dressing   saltwater   soil

8. What is the name of the concept that states that mass is neither created nor destroyed?

    _____

9. Which of these are processes that shape the surface of the Earth?

    migration   weathering   rock cycle   evaporation

10. Complete the chart by filling in the names of each phase of matter and at least one example of a substance that represents each phase of matter.

State of Matter	Description	Particles	Examples
A)	rigid	tightly packed, move less freely	
B)	flowing	farther apart, able to slide past one another	
C)	expanding	move very freely, lots of empty space between them, break apart easily	

205

Lesson #103

Series and Parallel Circuits

A **series circuit** is an electrical circuit that has only one pathway. Electricity flows from the source, through the conductor, through an appliance, and back to the source. If the path is interrupted in any way, the electricity will not flow through a series circuit. Series circuits are not very practical for use where a lot of electricity is needed. That's because one interruption, such as a switch that is turned off, will turn off the flow of electricity to the whole system or building. Also, as electricity travels along a series circuit, it becomes weaker and weaker the farther away it gets from the power source.

A better type of circuit for homes, schools, and businesses is a **parallel circuit**. You may know that parallel lines are side by side but not intersecting. This will help you understand how a parallel series works. The electric currents in a parallel circuit all have separate paths even though the currents may be coming from a common source. If one circuit is broken or interrupted, the currents will still flow along all the other paths as long as the source is still active. This means you can turn off the TV in the living room and the refrigerator will continue running in the kitchen. Your neighbors may have all their lights turned off, but you will still be able to use your lights. Another difference between a series circuit and a parallel circuit is that the electrical current in a parallel circuit is not weakened by use in its other lines. Each circuit has full power from the source even if there are very many appliances.

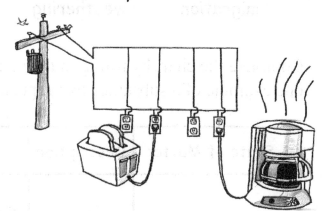

1. What is one distinction between a series circuit and a parallel circuit?

   _____

   _____

2. A circuit in which the current of electricity has multiple paths on which to flow is called a _____ circuit.

       insulated   switch   magnified   parallel

Simple Solutions© Science   Level 5

3. Which of these would interrupt the power supply in a series circuit?

   A) using a conductor   C) turning a switch on
   B) turning off a switch   D) using an appliance

4. The smallest unit of an element is an _____.

5. List three different types of fossil fuels.

   _____   _____   _____

6. Three of these organisms have something in common. Which one doesn't belong?

       oak tree   tomato plant   mushroom   grass

7. Explain why the one you chose in item 6 doesn't belong with the other three.

   _____

8. Which of these is an example of how organisms cause change in their environment?

   A) an earthquake   C) a flood
   B) plant roots breaking up rock   D) the movement of the tides

9. Atmospheric air pressure **decreases** _____.

   A) the closer you get to Earth's surface
   B) the higher up you go
   C) the more the Earth spins
   D) because of ultraviolet rays

10. How do decomposers benefit plants?

    A) Decomposers prevent flooding.
    B) Decomposers kill weeds.
    C) Decomposers put nutrients into the soil.
    D) None of these

Lesson #104

1. Why are electrical wires and extension cords usually covered with plastic?

   A) The plastic covering makes the wires neater.
   B) Plastic provides an additional layer of insulation.
   C) The covering is colorful and easy to see.
   D) Plastic keeps the wires from cooling too quickly.

2. Electricity is produced by the movement of _____.

3. A _____ opens and closes an electric circuit.

   battery   switch   conductor   current

4. Tina claims that sharpening a pencil is an example of a physical change. Is she correct? Why or why not?

   A) Tina is not correct because sharpening a pencil creates friction and that creates heat.
   B) Tina is correct because sharpening changes the form of the pencil but not its substance (what the pencil is made of).
   C) Tina is not correct because sharpening the pencil changes everything about it.
   D) Tina is correct because some of the matter that makes up the pencil is destroyed during sharpening.

5. Organic garbage, wood, crops, and manure are examples of _____.

   fossil fuels   biomass   polymers   edible produce

6. Which is an example of how organisms cause change in an environment?

   A) a construction crew building a shopping mall
   B) scavengers eating a dead carcass
   C) both A and B
   D) none of these

Simple Solutions© Science Level 5

7 – 10. Draw and label a diagram of the rock cycle. (See Lesson #31 or #56.) Use these terms:

sedimentary	igneous	metamorphic
weathering	sediment	compacting & cementing
heat & pressure	magma	melting & cooling

Lesson #105

Magnetism

Magnetism is a force. It is an **attraction** (pulling toward) or **repulsion** (pushing away) of magnetic materials. What causes this "attracting" and "repelling?" It is *the arrangement of electrons* in the atoms that make up **magnets** and magnetic materials. Iron and steel are the most common magnetic materials, and magnets will attract items made with iron, nickel, or cobalt. Every magnet is surrounded by a magnetic field. A magnet has two poles: a *north-seeking pole* and a *south-seeking pole*. These poles are where the magnetic field is the strongest. Often you will see a label – N or S – on the poles of a magnet. If you have two magnets, you will be able to identify the poles by holding them near each other. Like poles will repel; opposite poles will attract. If the two magnets pull toward each other, you know that one is the N pole and the other is the S pole. If they resist each other the two poles are the same.

Magnetism is a force that is very useful in everyday life. Magnets are used in compasses, motors, generators, builder's tools, scrap metal sorters, telephones, computers, doorbells, and all kinds of toys. Some magnets are very strong; others are very weak, and these differently powered magnets have different uses. Very powerful magnets may be destructive and dangerous. They can damage equipment like computers, and you can get a nasty pinch if your finger gets wedged between a strong magnet and a piece of metal!

1. The space around a magnet that holds a force is called the _____.

    power shield   force field   resistance   magnetic field

2. Which of these things are magnetic? Choose all that are.

    metal paper clips   plastic cups   magnets   ceramic   cars

3. What makes objects magnetic?

    _____

    _____

Simple Solutions© Science   Level 5

4. Look at each pair of magnets and decide how they will behave. Then write **attract**, **repel**, or **neither** next to each pair.

   A) _____   [N   S] [N   S]

   B) _____   [N   S] [S   N]

   C) _____   [S   N] [N   S]

5. Lightning is an example of _____ electricity.

       current   static   magnetic   hydro

6. Which of these is a way to be energy **efficient**?
   A) Using paper plates and napkins.
   B) Insulating windows to prevent heat loss.
   C) Getting up early in the morning.
   D) Watching the news before school in the morning.

7. Why do a cactus, a fern, and a tree belong to the *same kingdom*?
   A) They all grow outdoors.
   B) They all make their own food.
   C) They all die in cold weather.
   D) They all need water to survive.

8. Warm air can hold (more / less) water vapor than cold air.

9. Ice will become liquid when it reaches its _____ point.

       boiling   melting   condensation   freezing

10. True or False?

    _____ A chemical change does not create new matter.

Lesson #106

1. Put a check next to three ways of being more **energy efficient**.
    A)____ weather proofing doors/windows
    B)____ doing homework after dark
    C)____ using energy efficient appliances
    D)____ filling the dishwasher completely before running it
    E)____ using a pencil instead of a pen
    F)____ carpooling

2. What are the three states of matter?
    _____ _____ _____

3. Most of the ozone is within the _____.
    thermosphere   mesosphere   stratosphere   troposphere

4. Which of these is not a good conductor of electricity?
    aluminum   steel   copper   fiberglass

Look at the diagrams of circuits shown below.

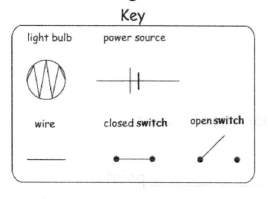
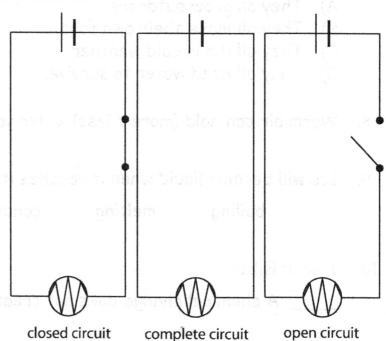

closed circuit   complete circuit   open circuit

5. Which of these will cause the bulb in the **complete circuit** to shine brighter?

   A) adding a switch
   B) adding more bulbs
   C) adding more power
   D) adding more wire

6. Which circuit will not allow the bulb to light up? _____

7. Which of these will cause the bulb in the **closed circuit** to become dimmer?

   A) adding more wire
   B) opening the switch
   C) adding more batteries
   D) all of the above

8 – 10. Complete the chart by naming the type of heat transfer that is described: **conduction**, **convention**, or **radiation**. Then give an example of each.

Type of Heat Transfer	Description	Example
	This transfer occurs when a mass of liquid or gas rises and falls. The energy must travel through gas or liquid.	
	This transfer occurs when a heat source comes into contact with a mass that is cooler. Objects must be touching.	
	This transfer occurs as energy travels in waves. It does not need to pass through any type of matter.	

Lesson #107

Sound Energy

Sound and light are both forms of energy that travel in waves. **Sound** is created by **vibrations** (back and forth movements of matter). You can probably think of many things that create sound vibrations: musical instruments, vocal chords, car engines, pencil tapping, and thunderstorms, just to name a few. Sounds can be loud or soft, low-pitched or high-pitched. Sound has volume, pitch, and frequency.

A sound's loudness is called **volume**, and the more energy a sound has, the greater its volume will be. Very loud sounds such as the blasts of fireworks or the pounding of heavy machinery can be painful or damaging to the human ear. People in jobs that are very noisy may wear earplugs or other protective gear because it is important to protect the ears from these high-energy, loud sounds. A sound's **pitch** is how high or low it is, and its **frequency** is the number of vibrations per second. The higher the frequency of sound waves, the higher the pitch will be. The lower the frequency, the lower the pitch will be.

Sound travels through matter in waves that move out in all directions. As sound waves move through air, the air particles vibrate. The sound keeps traveling until its energy runs out. If sound waves bump into something hard, they may bounce off, creating an echo. Or, sound waves may bump into something soft that will absorb the sound energy. Sound can travel through any kind of **medium** (matter that sound waves pass through), and sound waves move through different mediums at different speeds. Sound waves move faster through water than through air; they move through solids like steel and granite at even faster speeds. Remember, sound must have a medium to move through. If there is no air or water – like in outer space – sound does not travel.

1. The more energy a sound has the greater its _____ will be.

    distance   pitch   volume   level

2. Sound waves travel through different **mediums** at different speeds. Which of these is a medium for sound waves?

    air   water   steel   granite   all of these

3. When sound waves bump into matter, the sound may bounce off, creating a(n) _____.

4. The cello and the violin are two instruments that look similar. But the violin is much smaller than the cello and has much shorter strings. These smaller strings will vibrate faster than the long strings of the cello. Which of these is true?

   A) The violin will produce sounds with a higher frequency and pitch than the cello.
   B) The violin will produce sound with a lower frequency and pitch than the cello.
   C) The violin's notes will have higher frequency but lower pitch than those of the cello.
   D) The violin and the cello will produce sounds with equal frequency and pitch.

5. What are the six types of simple machines?

   _____

6. What do these words have in common?   weathering   erosion   uplift

   A) They name processes in the water cycle.
   B) They name severe weather events.
   C) They name processes that shape the surface of the Earth.
   D) They are processes that happen beneath the surface of Earth.

7. A _____ is the basic unit of life.

   consumer   producer   atom   seed   cell

8. Where will air pressure be greatest?

   A) at higher altitudes
   B) at lower altitudes
   C) air pressure is the same everywhere

9. Electricity at rest is called _____ electricity.

10. In an atom, when the number of protons and electrons is the same, the atom is _____.

    neutral   unstable   positively charged   negatively charged

Lesson #108

1. Which form of energy is created when something vibrates?

    wind   biomass   light   sound

2. Nicholas has a question: What causes mold to grow on bread? He wants to design an experiment. What should Nicholas do first?

    A) Get some bread and leave it on a table for several days.
    B) Do some research and write a hypothesis.
    C) Write a list of procedures for an experiment.
    D) Gather some materials like bread, water, and juice.

3. A renewable energy resource has an endless supply if it is protected. Which of these are renewable energy resources?

    waterpower   coal   wind   natural gas

4. Which group lists plants that store food in their roots?

    A) apple trees, corn plants, blueberry bushes
    B) onions, potatoes, carrots
    C) cacti, mushrooms, ferns
    D) fir trees, grass, dandelions

5. Which of these weather conditions show that the relative humidity has come close to 100%?

    dry heat   high winds   fog   sunny skies

6. What is a distinction between a predator and a scavenger? (See Lesson #12.)

    _____

7. The landform pictured here was probably created by which process?

    A) movement of water or glaciers
    B) shifting of tectonic plates
    C) an earthquake
    D) activity of prehistoric animals

8 – 10. Draw a diagram of each of the following: a complete circuit, an open circuit with a switch, and a closed circuit with a switch. Use the symbols shown below. (See Lesson #106.)

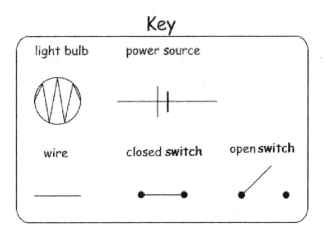

Lesson #109

Light Energy

Light is another form of energy that travels in waves. Like sound waves, light waves travel until they run out of energy. Light waves spread out in all directions and bump into things, also. Although sound needs to travel through a *medium* like air, water, or solids, light waves can travel through empty space. When light waves bump into objects, some of their energy is absorbed, and some of the energy bounces off of the object. Light bouncing off a shiny or smooth surface is **reflection**. Reflection allows you to see yourself or other things in a mirror or a window pane. Some objects, like clear water, glass or plastic, allow light to pass through them. These "see-through" objects are called **transparent**. Other objects like tinted glass, sheer curtains, and waxed paper allow only a little light to pass through. These materials are called **translucent**. Objects that do not allow any light to pass through are called **opaque**. Dark curtains, plaster walls, and most solids are opaque.

As light moves through one transparent substance into another, it bends. This bending is called **refraction**. A common example is light passing from air to water. If you put a pencil in a clear glass of water, it will look like the pencil is bent. This is because of refraction.

1. Objects that you cannot see through are called _____.

    opaque   transparent   translucent

2. You are able to see your own image in a mirror because of_____.

    translucence   refraction   reflection   transparency

This pool at the Taj Mahal reflects light from the sun in a pattern that creates a "mirror image" of the structures.

3. An experiment tests a _____.

    constant   communication   hypothesis   conclusion

Simple Solutions® Science   Level 5

4. Fill in the blanks, using these terms: systems   tissues   cells   organs

   _____ work together to form tissues.

   _____ work together to form _____.

   Organs work together to form _____.

5. When the number of protons is greater than the number of electrons in an atom, the atom is _____.

   neutral   unstable   positively charged   negatively charged

6. What causes prevailing winds?
   A) cold air above the poles moves toward the equator
   B) warm air at the equator rises and moves toward the poles
   C) both A and B
   D) water vapor forms as water evaporates from oceans

7. Which of these is a renewable resource?

   waterpower   petroleum   natural gas   coal

8. _____ is the amount of space that matter takes up.

9. An example of converting chemical energy to thermal energy is _____.

   fire   water   dry ice   turbines

10. Which of these would absorb sound the best?

    rocky cliffs   concrete walls   foam insulation   steel doors

As it moves from the air to the shiny part of the building, light is refracted, changing the angle at which you see the beam.

Lesson #110

1. Take a look at these three glass containers.

   Which container is made of **transparent** glass? _____

   Which container is made of **translucent** glass? _____

   Which container is made of **opaque** glass? _____

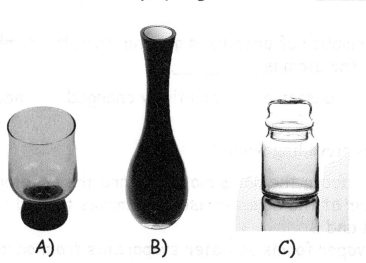

   A)   B)   C)

2. When has an ecosystem reached its carrying capacity?

   A) when the ecosystem can no longer meet the needs of organisms for air, food, water, and/or shelter
   B) when other organisms move in and decide to take over
   C) when the seasons change in that ecosystem
   D) when the populations decide that the area is full

3. About how long does it take for the moon to go through its phases?

   365 days　　　21 days　　　24 hours　　　30 days　　　7 days

4. An atom is **neutral** when its number of _____.

   A) protons is greater than the number of electrons
   B) electrons is greater than the number of protons
   C) neutrons is zero
   D) protons is equal to the number of electrons

5. In an element, all the _____ are the same.

   minerals　　　atoms　　　resources　　　cells

Simple Solutions© Science   Level 5

6. What are the three states of matter?

   _____   _____   _____

7. What type of energy is generated by vast quantities of water moving rapidly?

   A) waterfalls
   B) hydroelectric power
   C) water slides
   D) geothermal power

8. What do these terms have in common?

   volcanic eruptions   movement of glaciers   earthquakes   plant growth

   A) They name natural disasters.
   B) They name processes or events that shape the Earth's surface.
   C) They name events that occurred millions of years ago but do not occur today.
   D) They name events that affect weather patterns.

9 – 10. Complete the Venn diagram below. Use these terms:

   cell membrane   cell wall   chloroplasts   cytoplasm
         mitochondria   nucleus   vacuole

   Plant Cell   Animal Cell

Lesson #111

1. In which of these ways do decomposers help an ecosystem?

   A) killing off invasive species
   B) regulating the amount of water in soil
   C) putting vital nutrients back into the soil
   D) making food during photosynthesis

   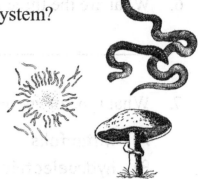

2. The **nucleus** of an atom contains smaller particles called

   _____ and _____.

3. Particles spinning in an orbit around the nucleus are called _____.

4. When molecules move faster and faster, what is the effect on temperature?

   A) Temperature increases.   C) Temperature remains the same.
   B) Temperature decreases.   D) Temperature goes up and down.

During science class, Miss Jasmine filled a bowl with lukewarm water and placed it on a lab table. She added hydrogen peroxide and quick-rising yeast to the water. As she stirred, the mixture started to bubble; the sides of the bowl felt warm to the touch.

5. What type of change – if any – occurred when the hydrogen peroxide and yeast were mixed in with water in the bowl?

   physical   chemical   atmospheric   no change

6. Miss Jasmine measured the temperature of the water when she first placed the bowl on the lab table. Then she measured the temperature of the mixture after adding the hydrogen peroxide and yeast, and the temperature was several degrees higher. What is the best explanation for this change in temperature?

   A) A chemical reaction produced heat energy.
   B) Heat was transferred from the table to the bowl by conduction.
   C) The ingredients that were added to the water were very hot.
   D) Stirring produced enough kinetic energy to raise the temperature of the mixture.

7. What was the *most likely* purpose of this activity?

   A) Miss Jasmine was following a recipe.
   B) Miss Jasmine was performing a demonstration.
   C) Miss Jasmine was building a model.
   D) Miss Jasmine was getting ready to bake some bread.

8. Give the name of the curly, wispy clouds shown in this photo.

   _____

9. Three of these terms have something in common. Which one doesn't belong?

       trees   water   air   natural gas

10. Look at the illustration of a straw in a glass of juice. Which property of light makes the straw look like it is bent or disconnected?

    A) reflection
    B) refraction
    C) absorption
    D) energy

Lesson #112

1. Which organism belongs to the same **kingdom** as snakes?

    rabbit   bird   worm   all of these

2. To test a hypothesis, a scientist conducts _____.

    a microscope   electricity   a theory   an experiment

3. This dolly makes it easier to move heavy boxes from one room to another. Which two simple machines are combined in a dolly?

    A) pulley & lever
    B) wheel-and-axle & screw
    C) lever and wheel-and-axle
    D) wedge and pulley

4. According to the Law of Conservation of Matter, mass is neither _____, nor _____.

5. Salamanders and frogs belong to which group?

    reptiles   birds   amphibians   mammals

6. Which of these is not a safe practice?

    A) flying a kite far away from power lines
    B) shutting off lights when they are not in use
    C) running an electrical extension cord through water
    D) playing volleyball in a swimming pool

7. What is the name for fuels that are from plants and animals that lived hundreds of millions of years ago?

    pesticides   solar power   prehistoric power   fossil fuels

A sound's **pitch** is how high or low it is. **Frequency** is the number of vibrations per second. The faster the vibrations, the higher the frequency of sound waves, and the higher the pitch will be. The slower the vibrations, the lower the frequency, and the lower the pitch will be. Musical instruments that are smaller vibrate faster than larger instruments.

8. Which instrument has a higher pitch?

   _____

   tuba   trumpet

9. Any object that has magnetic properties is called a(n) _____.

   rock   mineral   magnet   electron

10. The following are examples of physical and chemical changes. Sort the items into two lists.

    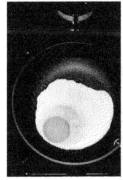

    souring milk   breaking glass
    raking leaves   rusting metal
    juicing a lemon   frying an egg

Chemical Change	Physical Change

Lesson #113

1. Which of the following are *abiotic* factors that can cause a change in an environment?

   metamorphosis   landslide   hibernation   drought

2. Which of these is a **mixture** of two or more solids?

   lemonade   sand   shaving cream   paint

3. Why is it important to have plenty of trees and other green plants in any ecosystem?

   A) The roots of trees soak up lots of extra water.
   B) Many kinds of fruit grow on trees.
   C) Trees absorb carbon dioxide and release oxygen.
   D) Trees grow from seedlings.

4 – 5. Use the phrase **high pressure** or **low pressure** to fill in each blank.

   Winds blow from areas of _____ to areas of _____.

6. Which type of simple machine is pictured here?

   _____

7. Which instrument has the <u>lowest</u> pitch? _____

snare drum

bongos

kettle drum

8. Three of these terms have something in common. Which one doesn't belong?

   metamorphic   igneous   sedimentary   lava

9. Explain why the one you chose in item 8 doesn't belong.

   _____

   _____

Look at the diagrams below. In a parallel circuit, the current flows separately from the source (battery) and through the appliance (bulb). An equal amount of energy is delivered to each bulb. A **parallel circuit** has branches, whereas a **series circuit** has only the one loop. Therefore, the current in a parallel circuit will not be interrupted or lessened by adding more light bulbs. Also, taking out one of the bulbs or flipping one of the switches will not interrupt power to the other bulbs in a parallel circuit.

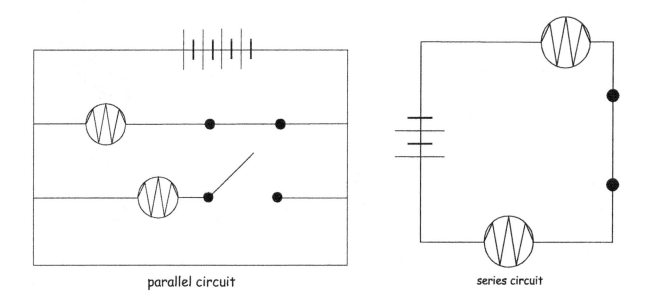

parallel circuit   series circuit

10. What is one of the advantages of a parallel circuit over a series circuit?
    A) A parallel circuit delivers electricity.
    B) As you add more bulbs the brightness will not weaken.
    C) One switch will turn everything off in a parallel circuit.
    D) A parallel circuit is open.

Lesson #114

Force

A **force** is a push or a pull that acts on an object. A force may cause the object to move, to stop, or to change direction. Some forces act by coming into contact with objects, like a golf club hitting a ball or the wind blowing against a curtain. Other forces, like gravity, magnetism, and electricity act at a distance, that is, without touching the object. All the things in our environment have forces acting on them all the time. Gravity is a force that is always acting on everything, and there may be other forces acting at the same time.

When you lean against a wall, you are applying a force. But the wall doesn't move and neither do you. That's because gravity and other forces are holding the wall in place, and those forces are equal to your push. When the forces acting on an object are **balanced**, the object's motion does not change. Another way to say this is that the forces are equal in strength, and they cancel each other out. When the forces acting on an object are **unbalanced**, its motion will change. If the object is still, it will move; if it is moving, the object may slow down, speed up, stop, or change direction.

1. What is a force?

   _____

2. If a car begins to slow down, the forces acting on it must be _____.

   A) balanced   C) unbalanced
   B) fragile   D) equally strong

3. What will cause a still object to move?

   a balanced force   an unbalanced force   neither   both

> When the forces acting on an object are balanced, there is no change in motion.

Simple Solutions© Science   Level 5

4. Which of these has the greatest average kinetic energy?

   ice cream   chocolate bars   fresh fruit   boiling milk

5. Matter is anything that _____.
   A) takes up space
   B) is made of atoms
   C) has mass
   D) all of the above

6. Put a check next to three ways of being more **energy efficient**.

   ___using long lasting light bulbs   ___planting trees
   ___recycling bottles and cans   ___insulating windows and doors
   ___using water-saving toilets   ___wrapping garbage in newspaper

7. Sometimes loose electrons move through matter. This creates _____.

   electrical currents   new atoms   new elements

8. The Periodic Table of Elements is a list of what?
   A) all the known elements
   B) prehistoric periods of the Earth
   C) scientists that have discovered atoms
   D) minerals and their properties

9. The condition of the atmosphere over long periods of time in a given place is _____.

   temperature   air pressure   climate   air quality index

10. What does a circuit need in order to be complete and to work properly?
    A) switch, insulator, and light bulb
    B) power source and conductor
    C) conductor and insulator
    D) switch and power source

> When the forces acting on an object are unbalanced, the object will move, stop or change direction.

Lesson #115

Friction and Buoyancy

When two objects come into contact with one another, there is always a certain degree of friction. **Friction** is a force that reduces motion by working against it. Even a very smooth surface – like a skating rink – has some friction, and you would be able to see it at a microscopic level. Other things that cause friction are very easy to see, like speed bumps in a parking lot. Friction is what allows you to stop your bike by applying the brakes. The brakes rub against your bicycle tires, and that slows or stops the motion. Friction always *works against* motion.

Buoyancy is another force; it effects objects that are in fluids, usually water. Buoyant force is the force of water pushing up on an object and keeping it afloat. When a raft is floating on a pond, the raft moves some of the water out of place. The buoyant force of the raft is equal to the weight of the water that is pushed aside by the raft itself. Like friction, buoyant force acts *in opposition* to another force – gravity. If the weight of an object is greater than the weight of the water it displaces, there will not be enough buoyant force, and the object will sink.

1. Which **forces** cause the sled to move?

    A) gravity and the adult pushing the sled
    B) buoyancy and friction
    C) friction and velocity
    D) the steering bar and gravity

2. What **force** allows a sailboat to stay afloat?

        buoyancy   gravity   friction   wind

3. Which of these is a mixture of two or more gases?

        soda pop   humus   air   petroleum

4. The innermost layer of the Earth is called the _____.

        core   mantle   crust   continents

5. Which layer of the atmosphere has the lowest density? (See Lesson #50.)

        troposphere   stratosphere   mesosphere   thermosphere

Simple Solutions© Science Level 5

6. Look at the diagram of a circuit. What will happen to the bulb in this circuit?

    A) The bulb will light but only for a short time.
    B) The bulb will light, but it will flicker.
    C) The bulb will light and remain lit.
    D) The bulb will not light up.

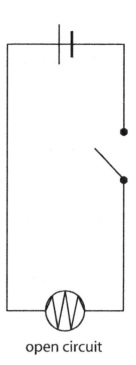

open circuit

7. Look at the diagram of the layers of Earth's crust. Which layer would most likely contain the oldest fossils?

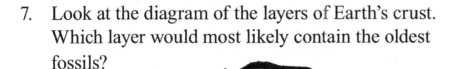

8. **True or False?**

    _____ Renewable resources like water may become unusable due to pollution.

    _____ If a forest is completely cut and removed, it will automatically grow back.

    _____ Fish are a renewable resource, so humans can use up as many fish as they want without ever running out.

9. The food you eat combines with oxygen during a chemical reaction that releases heat and energy. What is this chemical reaction called?

    recreation   aspiration   digestion   convection

10. _____ is a measure of average kinetic energy.

    Weight   Mass   Speed   Temperature

Lesson #116

1. Which kind of change does not produce a different substance?

    physical   chemical   both   neither

2. Friction is an opposing force acting against the motion of an object. What provides friction against the motion of the three-wheeler in this photo?

    A) the little boy and the wind

    B) wind and gravity

    C) grass and gravity

    D) the wheels and the boy

3. An atom that is **neutral** has an equal number of _____.

    protons and neutrons   protons and electrons   neutrons and electrons

Complete the following sentences using terms from the word bank.

| photosynthesis | nutrients | consumers |
| decomposers | scavengers | producers |

4. _____ are plants that make their own food through photosynthesis.

5. _____ are animals that eat plants. When an animal dies, microorganisms called

6. _____ break down the animal's body and return

7. _____ to the soil.

232

8. Choose the word that completes the sentence correctly.

   The melting point of a substance will always be (higher / lower) than its boiling point.

9. Label this diagram of the water cycle.

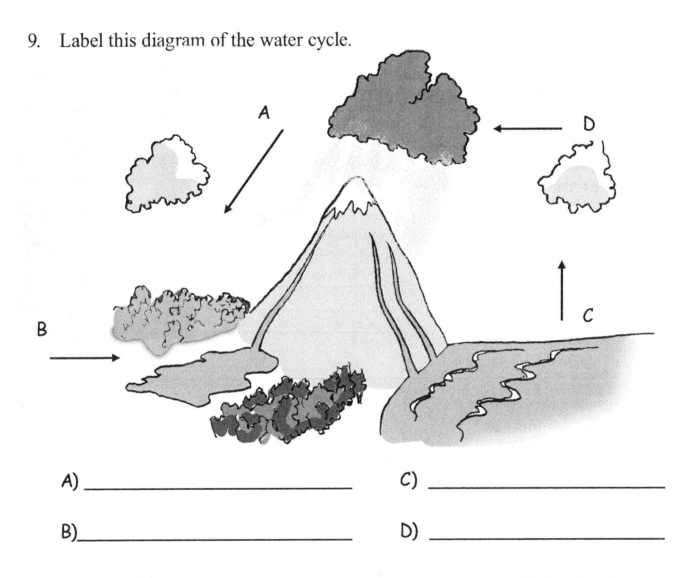

   A) _____   C) _____

   B) _____   D) _____

10. Which state of matter has definite mass, definite volume, and definite shape?

    gas   solid   liquid   all three

Simple Solutions® Science   Level 5

Lesson #117

1. An atom will have a positive charge when _____

    A) it has more electrons than protons.
    B) it has more protons than electrons.
    C) it has no neutrons.
    D) it has the same number of protons, neutrons, and electrons.

2. Janet and Phyllis were making S'Mores. Janet toasted the marshmallows until they turned brown, and Phyllis melted the chocolate by putting the hot marshmallows on top of chocolate squares and graham crackers. Both girls claimed that they were creating a chemical change.

    Are they correct? Write your explanation below.

    _____

    _____

    _____

3. What is **biomass**?

    A) oil, natural gas, and coal
    B) organic material that can be burned to release energy
    C) energy generated from wind turbines
    D) heat energy from the center of the Earth

4. The condition of the atmosphere at any given time in a given place is _____.

        density   temperature   humidity   weather

5. Every _____ has a north-seeking pole and a south-seeking pole.

        valley   landform   atom   magnet

6. Which type of material allows some light to pass through it?

        opaque   translucent   solid   black

7. When the number of electrons is greater than the number of protons in an atom, the atom is _____.

   neutral   unstable   positively charged   negatively charged

8 – 10. Show what you know about simple machines. Complete the chart by filling in the missing names or descriptions.

Machine	Description	Illustration
A)	an inclined plane spiraled around a post used to fasten or hold things together	bottle top
inclined plane	B)	
C)	uses grooved wheels and ropes to raise and lower things	
wedge	D)	
E)	bar that pivots on a fulcrum to lift or move heavy loads	
wheel-and-axle	F)	

Lesson #118

Newton's First Law of Motion

Everything in our environment is always in motion. Particles in atoms are vibrating. Earth is rotating on its axis and revolving around the sun; air is moving, and other things that we can see are also in motion. Remember, a **force** is needed to get an object to move, change speeds, to stop, or to change direction. This can be a **contact force** (matter touching matter), or it can be a force that acts from a distance such as gravity, magnetism, or electricity. Newton's Laws of Motion explain why things move and how movement changes. There are three of these laws.

Newton's first law says, **"An object at rest tends to stay at rest, and an object in motion tends to stay in motion at the same speed and in the same direction unless it is acted upon by an unbalanced force."** That's a lot to remember, so a short way to say it is this: "Objects have inertia." **Inertia** (in er' shuh) means an object will not change its state of motion unless it experiences an outside force. Objects that are still will remain still, and objects that are moving will not change their motion unless an outside force makes them move. It will not change direction, slow down, or stop moving unless a force makes it change direction, slow down, or stop.

Imagine a soccer ball resting on a field. The ball will not move unless some force acts on it. A strong wind could move the ball a little, or someone could bump into it or pick it up. Let's say you kick the ball, and it goes flying through the air. Once it is moving, Newton's Law says the ball will continue going in the same direction at the same speed unless a force acts upon it. As the ball is sailing through the air, it may be turned in a different direction by the wind. **Gravity** will eventually bring the ball to the ground. Once it's on the ground, the ball may roll a little, and **friction** is another force that will come into play. Contact with the grass provides friction, and that will slow the ball. It will eventually come back to stillness.

1. Name three **contact forces** that act on the soccer ball in the example described above.

   _____

2. Some of the forces acting on a soccer ball are the kicker's foot and friction from the wind or the grass. Name another.

   _____

Simple Solutions© Science   Level 5

3. The tendency of an object in motion to stay in motion or an object at rest to stay at rest is _____.

    force   gravity   inertia   motion

4. One way that plant and animal cells are alike is that both have _____.

    chloroplasts   a nucleus   a cell wall   photosynthesis

5. In the scientific method, which comes first?

    question   conclusion   data chart   report

6. What are the three states of matter?

    _____   _____   _____

7. Which simple machine is made with a rope, chain, or belt wrapped around a curved wheel?

    screw   inclined plane   pulley   wheel-and-axle

8. A device that measures the speed of wind is _____.

    thermometer   meter stick   anemometer   wind sock

9. What does *metamorphic* mean?

    water to land   changing form   from the Earth   heat and pressure

10. Label the parts of this circuit: **power source, conductor, appliance.**

Lesson #119

Frame of Reference, Speed, and Velocity

How do you know something is moving? You know because other things are not moving. When you are going up an escalator, you notice that everything else seems to be going down. If you were in an airplane, you might peer through a window and see objects that appear to be shrinking. The position that you are in as you experience motion is called a **frame of reference**. If you are watching a baseball fly across a field, your frame of reference would be the field and everything around it: fence, people, trees, bases, and stadium seats.

Speed is how far an object travels in a certain amount of time. For example, if you can bike 36 miles in three hours, your speed is 12 mph. Speed is calculated by dividing the distance traveled by the amount of time it takes to travel. Thirty-six miles divided by 3 (hours) equals 12. **Velocity** is speed in a specific direction. If you are riding your bike north at 12 mph, that is your velocity. When you change direction or change speed, you change your velocity. Velocity can only be measured if you know both the speed and the direction in which an object is traveling.

1. Students are riding on a school bus and looking out the windows. Buildings, parked cars, people waiting to cross the street provide a _____ for students on the bus.

    frame of reference   speed and velocity   specific direction

2. The distance an object travels in a certain amount of time is its _____.

    velocity   direction   speed   inertia

3. Velocity is a measure of both _____

    and _____.

4. Which of the following provides a <u>contact</u> force?

    wind   gravity   magnetism   electricity

5. An object at rest will remain at rest, and an object in motion will remain in motion at the same speed and in the same direction unless _____ acts upon the object.

    a balanced force   an unbalanced force

Simple Solutions© Science Level 5

6. Newton's first law of motion is mostly about _____.

   gravity   speed   friction   inertia

7. Label this diagram of an atom. Use these words: nucleus, proton, neutron, electron.

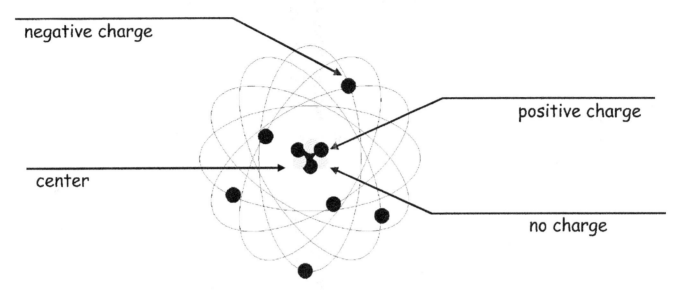

negative charge

positive charge

center

no charge

8. Organisms in an ecosystem have needs such as air, food, and water. An ecosystem has reached or exceeded its _____ when it can no longer meet these needs.

   production level   liability   population   carrying capacity

9. The distance traveled divided by the time it takes to travel is _____.

   speed   velocity   acceleration   frame of reference

10. The photo shows that a _____ has no definite shape or volume.

    gas   liquid   solid   all three

Lesson #120

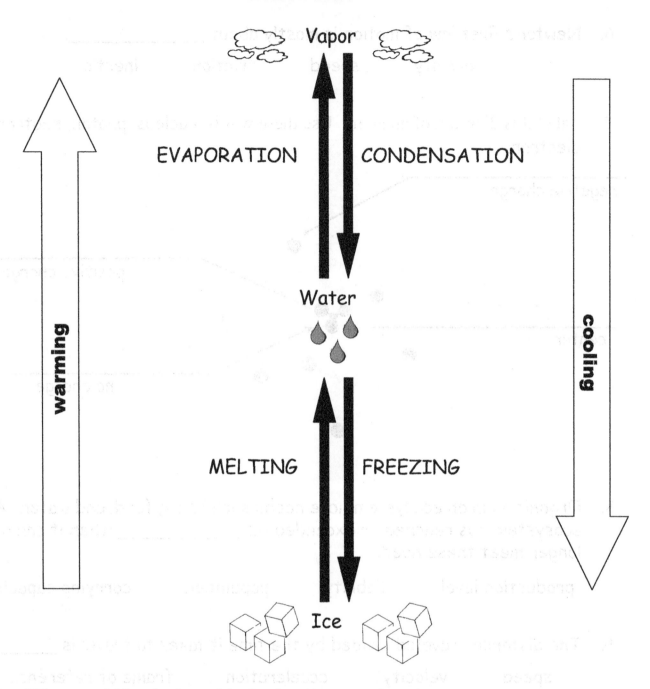

1. What would be the best title for the diagram above?

   A) The Nitrogen Cycle
   B) Phase Changes of Water
   C) Melting Point and Boiling Point
   D) Physical and Chemical Change

2. What causes water to move from one phase to another?

   atmospheric pressure   Earth's rotation   transfer of heat energy

3. Which change occurs during condensation?

   vapor to solid   liquid to solid   vapor to liquid   solid to vapor

Simple Solutions© Science   Level 5

4. When the number of electrons is greater than the number of protons in an atom, the atom is _____.

   neutral   unstable   positively charged   negatively charged

5. Sam throws a basketball up into the air. As it lands on the gym floor, the ball bounces several times, then rolls a little and comes to a stop. What causes the ball to stop moving?
   A) friction and gravity
   B) bouncing, noise, and air flow
   C) gravitational potential energy
   D) wind, sound, and vibration

6. Which of the following tells the **velocity** of an object?

   90 mph   16 kilometers   20 mph east   6 rpm

7. In which layer of the atmosphere do airplanes fly?

   _____

8. Three of these organisms have something in common. Which one doesn't belong?

   bat   dolphin   turtle   giraffe

9. Explain why the one you chose in item 8 doesn't belong with the other three.

   _____

   _____

10. Which word would you use to explain why toy ducks float on top of water instead of sinking to the bottom?
    A) buoyancy   C) friction
    B) magnetism   D) static electricity

Lesson #121

Norbert Rillieux, Scientist, Engineer, & Inventor (1806 – 1894)

Norbert Rillieux was born in New Orleans, Louisiana in 1806. Many sources say he was born a slave or that his mother was a slave. But Norbert was raised as a "freeman" even though he lived at a time when most African Americans in the South were slaves. Norbert's mother worked for the owner of a sugar plantation, and they were dependent upon the wealthy whites for food and shelter. When Norbert was old enough, the plantation owner decided to send him to a Catholic school instead of making him work in the fields. Norbert was very bright, and he excelled at his school work. But he also saw the slaves at work and observed how sugar was harvested. He understood the back-breaking work that went into cutting the sugarcane. And he knew that the slaves had to build raging hot fires to process the sugar. The work was dangerous because slaves used giant ladles to pour searing hot liquid sugarcane from one container to another.

When Norbert was older, he was sent to study in Paris, France, and by the age of twenty-four, he became an instructor there. Norbert studied engineering, and he realized that the boiling point of a liquid could be lowered by lowering air pressure. He experimented with heating sugarcane juice in a vacuum. This made the process much easier and less wasteful. As a result, Norbert invented a new distillation process for making sugar. His invention made the refining of sugar less dangerous for workers. This process could save money, so plantation owners in the United States wanted to use it right away. Norbert Rillieux held patents for other inventions and did many kinds of work in his lifetime. But he is most remembered for the way he revolutionized the sugar industry.

Harvesting sugarcane and refining sugar was brutal, dangerous work for slaves in the 1800's.

Simple Solutions© Science   Level 5

1. What was one difference between Norbert Rillieux and other African American children of his time?
   A) Norbert was very bright.
   B) Norbert lived on a sugar plantation.
   C) Norbert was allowed to go to school.
   D) There was no difference between Norbert and other children.

2. Why was Norbert Rillieux's invention so important?
   A) It made sugar processing safer and more profitable.
   B) It made sugar processing less wasteful.
   C) Rillieux was one of the first African American inventors.
   D) all of the above

3. Rillieux knew that one way to lower the boiling point of a liquid is to lower the _____.

   air pressure   thermal energy   fire   amount of sugar cane

4. Making sugarcane into refined sugar (table sugar) requires a (physical / chemical) change.

5. Warm air has a (higher / lower) density than cold air.

6. The **melting point** of a substance will always be (higher / lower) than its **boiling point**.

7. Three of these terms have something in common. Which one doesn't belong?

   producer   decomposer   consumer   manager

8. Explain why the one you chose doesn't belong with the other three.

   _____

9. Rock is broken down into soil through a process called _____.

   weathering   migration   natural resources   conservation

10. Which organism is most likely to be at the bottom of a food chain?

    whale   squirrel   cow   raccoon   algae

Lesson #122

Newton's Second Law of Motion

Remember velocity is speed in a specific direction. Velocity changes whenever the speed or the direction changes. Let's say you start walking in a certain direction. Your muscles warm up and you begin to pick up the pace. Your speed increases. Your velocity is changing.

Acceleration measures how quickly the velocity is changing or the *rate* at which velocity *changes*. Think of two cars traveling at 35 mph. The driver of car #1 sees a stop sign ahead and taps the brakes. Car #1 gradually comes to a stop. The driver of car #2 sees a squirrel run into the street, slams on the brakes, and comes to a halt. Both cars went from 35 mph to 0 mph (stopped), but the velocity of car #2 changed much more quickly than that of car #1. Car #1's acceleration was smaller than car #2's because its velocity changed at a slower rate.

Newton's second law says, **"An object's acceleration is a result of both its mass and the amount of force applied to the object."** It takes more force to affect the acceleration of an object with greater mass; it takes less force to affect the acceleration of an object with smaller mass. Imagine a soccer ball and a bowling ball sitting on the floor of your classroom. You give the soccer ball a gentle kick and it rolls across the floor. Now, you give the bowling ball the same gentle kick. Ouch! The bowling ball may move, but it won't roll as far or as fast as the soccer ball. The bowling ball has a greater mass than the soccer ball, and will need a greater force in order to change its acceleration. The same is true of stopping an object or changing its direction. It takes more force to slow down or stop a heavier object and less force to slow or stop an object with a smaller mass.

1. According to **Newton's second law of motion**, *acceleration* is the result of both the _____ of an object and the amount of _____ applied to the object.

2. The rate at which velocity changes is _____.

3. A(n) _____ is required in order to move, stop or change the direction of an object.

        acceleration   velocity   object   force

Simple Solutions© Science Level 5

4. According to Newton's first law, a paper airplane that is launched into the air should keep going until something makes it stop. What is one force that would cause a paper airplane to stop sailing through the air?

_____

_____

5. The photo shows that a _____ keeps its shape and volume inside or outside of a container.

    gas   liquid   solid   all three

6. Which organism belongs to the same kingdom as trees?

    algae   mushrooms   cacti   birds

7. Look at the illustration at the right. Which property of light makes the polar bear appear distorted?

A) reflection

B) absorption

C) refraction

D) translucence

8. Which holds more moisture?   cool air   warm air

9. Three of these terms have something in common. Which one doesn't belong?

    hurricane   tornado   blizzard   earthquake

10. Explain why the one you chose doesn't belong with the other three.

_____

_____

Lesson #123

Newton's Third Law of Motion

When you exert a force on something, it is exerting a force on you. Lean against a wall; the wall is pushing against you. Set a plate on a table; the table is pushing against the plate. You cannot see evidence of the force in these examples, but think about what happens when you blow up a balloon but do not tie it. When you let go of the balloon, it sails across the room. Air is flowing out in one direction, and the balloon is thrust in the opposite direction. This will help you to understand Newton's third law of motion: **For every action, there is an equal and opposite reaction.**

1. According to **Newton's third law of motion**, all forces occur in pairs. The pair of forces are always _____.

    A) equal and opposite
    B) unequal and opposite
    C) equal and parallel
    D) opposite in force but equal in direction

2. Which of these explains what **Newton's third law of motion** has to do with how a rocket is launched?

    A) Gravity applies a force to the rocket, and this force along with the rocket's mass propels the rocket into the atmosphere.
    B) The rocket's engines supply a thrust that pushes on the ground; the ground pushes the rocket up with an equal and opposite force.
    C) A rocket in motion will stay in motion unless acted on by an outside force.
    D) Rocket fuel provides heat energy which is converted to motion energy at take-off.

3. **Newton's first law of motion** is also called the law of _____.

    gravity   acceleration   inertia   reciprocation

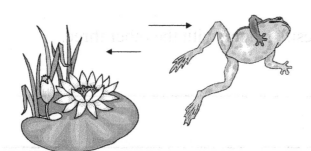

When a frog jumps off a lily pad, the frog moves in one direction, and the lily pad moves in the opposite direction.

4. Which of the following is an *abiotic* factor that can cause change in an environment?

    hibernation   photosynthesis   avalanche   migration

5. There are many different kinds of ecosystems. Which of these can be an ecosystem?

    ocean   swamp   forest   tundra   all of these

6. Look at the photo. The lemons are sinking, but the bubbles are floating to the top. Why?

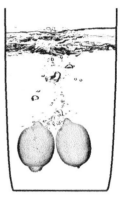

   A) There are no forces acting on the bubbles.
   B) The lemons have a lower density than the water.
   C) The air bubbles have a lower density than the water.
   D) The lemons have an aerodynamic shape.

7. The place where two different air masses meet and bump into each other is called a _____. This is where weather usually changes.

    border   latitude   longitude   front

8. Water will become vapor when it reaches its _____ point.

    melting   condensation   freezing   boiling

9 – 10. Label the parts of the plant cell and the animal cell.

Plant Cell

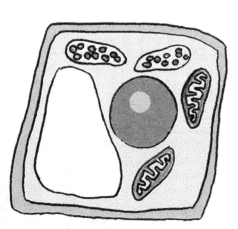

Animal Cell

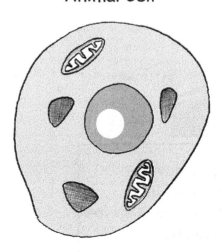

Lesson #124

1. How do you know that dissolving sugar in water is a <u>physical</u> change and not a chemical change?

   A) The solution is still sugar and water – no new substance is released.
   B) The sugar dissolves easily.
   C) More sugar will dissolve as the water is heated.
   D) Other substances can be added to the solution at any time.

2. How do you experience **Newton's first law of motion** while riding a skateboard?

   A) It is impossible to ride a skateboard uphill.
   B) The skateboard is made of wheels and a flat surface.
   C) Once the skateboard is going, you can continue forward for some time without pushing.
   D) Gravity keeps you on the ground while riding.

3. Air pressure in the troposphere allows living things to _____.

   A) stay grounded on Earth
   B) breathe the right mix of gases
   C) get a suntan
   D) be protected from ultraviolet rays

Two or more liquids may be combined to create a solution.

A solid may be dissolved in a liquid, making a solution.

A gas may be dissolved in a liquid; this is a type of solution.

Simple Solutions© Science   Level 5

4. Which pair of organisms is most closely related?

    protozoa/amoeba   mold/cactus   cactus/prairie dog

5. In our lifetimes, it is likely that _____ resources will be used up.

    renewable   nonrenewable   no

6. Warm air can hold (more / less) water vapor than cold air.

7. How should data be recorded during an experiment?
   A) Pay attention and try to remember everything that happens.
   B) Write everything down as soon as the experiment is over.
   C) Take notes during the experiment and organize the notes afterwards.
   D) Write a lab report before beginning the experiment.

8. What are the two ends of a magnet called?

   _____

9 – 10. Look at the drawing below. How does this illustration help to explain Newton's first law of motion? Write your explanation below.

Lesson #125

Use the diagram of a food web to answer the first three questions.

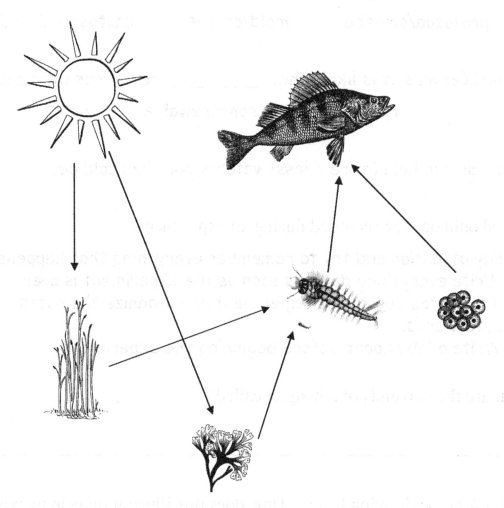

1. The diagram shows that all energy comes from _____.

   green plants   the sun   fish eggs   mosquito larva

2. The perch gets energy from the sun by _____.

   A) eating mosquito larva
   B) swimming near the surface of the water
   C) eating green plants
   D) breathing through its gills

3. In this food web, the producers are _____.

   A) mosquito larva and fish eggs
   B) algae and sea grass
   C) the sun and the perch

4. According to **Newton's third law of motion**, forces occur in _____.

    isolation   thirds   pairs   none of these

5. According to Newton's first law of motion, if an object is moving, the only thing that can stop it is _____.

    time   a force   brakes   air pressure

6 – 10. Sort the terms listed below into five different categories. Use each word only once.

hurricane	igneous	tsunami	condensation
soil	evaporation	wind power	precipitation
flooding	freezing	melting	thunderstorm
lava	blizzard	solar power	metamorphic
drought	mudslide	sediments	earthquake
tornado	avalanche	water power	sedimentary
biomass			

Phase Changes	Weather Events	The Rock Cycle	Renewable Energy Resources	Abiotic Factors that Affect an Environment

Lesson #126

1. Ramon used a flathead screwdriver to pry open a can of paint. The screwdriver was used as which kind of simple machine?

   screw   lever

   inclined plane   wedge

   wheel-and-axle   pulley

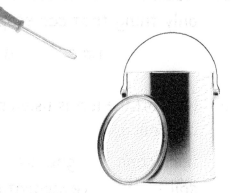

2. During digestion, acids in the stomach break down food particles so that the nutrients can be absorbed and transformed into energy. Why is digestion considered to be a chemical reaction?

   A) Molecules are rearranged.
   B) Matter is created.
   C) Matter is destroyed.
   D) All of the above occur.

3. Name an *abiotic* factor that can cause change in an environment.

   _____

4. What do these terms have in common?

   weathering   deposition   erosion   movement of glaciers

   A) They name processes that shape the Earth's surface.
   B) They name processes involving living things.
   C) They are part of the water cycle.
   D) They are processes that cause earthquakes.

5. When the number of protons and electrons is the same in an atom, the atom is _____.

   neutral   unstable   positively charged   negatively charged

Simple Solutions© Science — Level 5

6 – 10. Show what you know about the phases of matter. Complete the graphic organizer below with as much information as you can remember. How are the molecules arranged? How do they move? Does the phase have a definite shape, a definite volume? Give an example of each. (Lessons #63 – #65)

Phase	Particles	Shape	Volume	Example
Solid				
Liquid				
Gas				

Lesson #127

The Solar System

Our **solar system** is made up of a star that we call the sun and everything that travels around it. The eight planets that we know about and their moons, along with dwarf planets, comets, and asteroids all revolve in an orbit around the sun. An **orbit** is a path that a body in space follows around another body in space. The moon follows an orbit around the Earth, and the planets follow an orbit around the sun. The Earth not only revolves around the sun, it also rotates on its **axis** (an imaginary line through the center of Earth). It takes about 24 hours for the Earth to complete one full rotation; this rotation is how we measure the time in one day. It takes about 365 ¼ days for the Earth to complete one full revolution around the sun. Because of the way the Earth rotates and revolves, the Earth's hemispheres have opposite seasons. While the Northern Hemisphere is having winter, the Southern Hemisphere is having summer. It is spring in the Northern Hemisphere when it is fall in the Southern Hemisphere.

The sun is the biggest object and takes up the most space in the solar system. The sun is right in the middle of the solar system, and everything that travels around it moves in an **elliptical** (egg-shaped) orbit. The solar system is part of an even bigger system called the Milky Way Galaxy; the Milky Way is made up of over a hundred billion stars. And that's not all – there are billions of other galaxies in the universe as well!

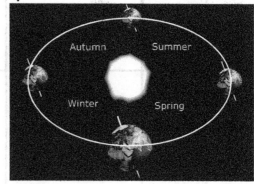

1. The biggest object in the solar system is the _____.

2. Why do some areas experience four seasons, each with different amounts of light and different temperatures?

    A) Sometimes the Earth is closer to the sun than at other times.
    B) The Earth both rotates on its axis and revolves around the sun.
    C) The sun burns at various temperatures at different times of the year.
    D) During the cooler seasons the moon is blocking the rays of the sun.

3. An _____ is a path that a body in space follows around another body in space.

4. How long does it take for the Earth to complete one full rotation? _____

5. Which of these is most like the shape of the orbit of planets around the sun?

A) 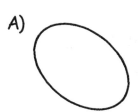 B) 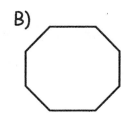 C) 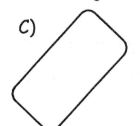 D)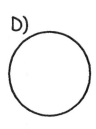

6. Match these examples of solutions.

   ___ a solid dissolved in a liquid   A) food coloring in water
   ___ a gas dissolved in a liquid   B) chocolate drink powder in milk
   ___ a liquid dissolved in a liquid   C) carbonated beverage

7. Why is it important to wear safety goggles when mixing isopropyl alcohol with food coloring?

   A) Food coloring may stain skin and clothing.
   B) Isopropyl alcohol can damage the eyes.
   C) Safety goggles make it easier to see and to measure liquids.
   D) There may be an explosion.

In a series circuit that is complete, the current flows from the source, through the conductor, through the appliance, and back to the source, making a complete loop.

8. What will happen if the switch is opened?

   A) Nothing will happen.
   B) The bulb will flicker.
   C) The light will go out.
   D) The bulb will become brighter.

9. What will happen if another battery is attached to the circuit?

   A) The bulb will flicker.   C) The bulb will become brighter.
   B) The light will go out.   D) Nothing will happen.

10. What will happen if two more light bulbs are attached to the circuit?

   A) The bulbs will become brighter.   C) The bulbs will go out.
   B) The bulbs will dim.   D) Nothing will happen.

Simple Solutions© Science Level 5

Lesson #128

1. In the solar system, the object with greatest of mass is _____.

   Earth   the asteroid belt   the sun   Mars

2. The Northern Hemisphere has the fall season while the Southern Hemisphere has _____.

   winter   spring   summer   fall

3. True or False?

   _____ An object with a larger mass requires greater force to accelerate than an object with a smaller mass.

   _____ According to Newton's Third Law, forces occur in pairs.

4. Why must a circuit be **complete** in order to work properly?
   A) A complete loop allows the electrical current to flow through.
   B) A complete circuit is neater.
   C) A circuit must always have a battery and a light bulb.
   D) A complete circuit will never run out of energy.

5. A) About how long does it take for the Earth to complete one full rotation on its axis?

   365 days   30 days   24 hours   60 minutes

   B) About how long does it take for the Earth to complete one full revolution around the sun?

   _____

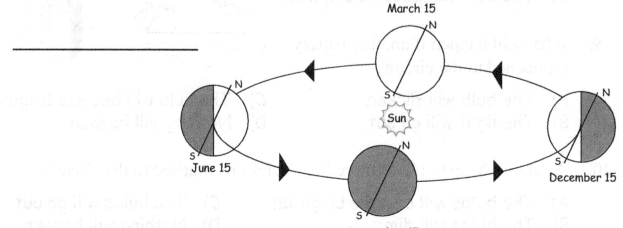

6. In a tug-of-war that neither team is winning, the forces are _____.

   balanced   unbalanced   strong   losing

7. What force allows this huge iceberg to float instead of sinking to the bottom of the ocean?

   A) inertia
   B) gravity
   C) buoyant force
   D) friction

8. Which simple machine makes it possible to move this chair across a room?

   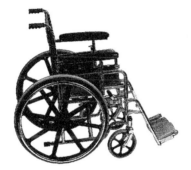

   _____

9. What are the three states of matter?

   _____

   _____

   _____

10. How do you know that **decomposition** is a chemical change and not a physical change?

    A) It is almost impossible for scientists to observe decomposers at work.
    B) Decomposers digest dead plants and animals.
    C) The decaying organic material can never be alive again.
    D) Decomposers are live organisms.

Simple Solutions© Science Level 5

Lesson #129

1. The following is a list of processes that cause physical or chemical changes. Sort the items and complete the list.

   shredding paper   decaying garbage   grilling a steak
   burning incense   melting ice cream   hammering a nail

   Physical Changes
   Matter changes in its form but does not change in its substance.

   _____

   _____

   _____

   Chemical Changes
   Atoms are rearranged and matter changes in substance.

   _____

   _____

   _____

2. Which <u>two</u> simple machines are used in a paint roller?
   A) lever   B) pulley   C) wedge
   D) inclined plane   E) wheel and axle

3. Match each weather instrument with the condition it measures.

   _____ thermometer   A) air pressure
   _____ barometer   B) wind speed
   _____ anemometer   C) temperature
   _____ hygrometer   D) humidity

4. Sometimes the atmosphere is called *a blanket of air surrounding the Earth*. Why is this description appropriate?
   A) There is plenty of electricity in the atmosphere.
   B) The atmosphere looks like a quilt.
   C) The atmosphere helps to keep the Earth warm.
   D) There are many dust particles in the atmosphere.

Simple Solutions© Science Level 5

5. Which of the following are ways to save energy in your classroom? Check all that apply.

   ____ turning off lights when no one is in the classroom

   ____ letting water run while you wash your hands

   ____ using both sides of a sheet of paper

   ____ using glue instead of paste in art class

6. An atom is neutral when _____.

   A) there are no neutrons present in the atom
   B) the number of protons is equal to the number of electrons
   C) the number of protons is greater than the number of electrons
   D) the number of electrons is greater than the number of protons

For each organism in the chart, tell which vertebrate group it belongs to and which type of consumer it is. Consider the animal's whole lifetime – it may start out as an herbivore and then eat other animals as an adult.

Vertebrate Groups: amphibian, bird, fish, mammal, reptile

Types of Consumers: herbivore, carnivore, omnivore

	Organism	Vertebrate Group	Type of Consumer
7.	Atlantic salmon		
8.	ostrich		
9.	grizzly bear		
10.	salamander		

259

Lesson #130

The Inner Planets

A **planet** is a large mass that has settled into a nearly spherical (round) shape and orbits a star. A planet has "cleared the neighborhood around its orbit." That means the planet is the main object in its orbit; it is not part of an asteroid belt, and it does not enter the orbit of any other object in space. Eight planets orbit the sun in our solar system. (There are also "dwarf planets" like Pluto, Ceres, and Eris.) Scientists have classified the planets into two groups: the inner planets and the outer planets. The four "inner" planets are the closest to the sun: Mercury, Venus, Earth, and Mars. These four planets are compact and rocky, but they are not similar in other ways.

Mercury is the closest planet to the sun. It does not have a life-supporting atmosphere, and its temperatures are always extremely hot or extremely cold. Temperatures on Mercury can range from -183°C to 427°C (-298°F to 800°F). **Venus** is the second planet from the sun and was once called Earth's "twin" since it is so close to Earth in size. However, the atmosphere of Venus is very dense and does not allow heat from the sun to be released. Temperatures on the surface of Venus are hot enough to melt metal! Venus' atmosphere has large amounts of carbon dioxide and sulfuric acid. The atmosphere of Venus is so dense that it would crush anyone who tried to stand on its surface. **Earth** is the third planet from the sun, and of course, it is the planet most familiar to us. Only Earth has the characteristics that make it **habitable** (able to support life) for humans and all of the other organisms we know. Nearly 70% of Earth's surface is covered by water, and its atmosphere is the perfect mix of nitrogen, oxygen, and other gases. Earth's distance from the sun and its atmosphere keep its average temperatures comfortable for Earthlings. The fourth inner ring planet is **Mars**, nicknamed the "Red Planet" because of its rusty colored surface. Mars has a very thin atmosphere composed mostly of carbon dioxide. The water on Mars is frozen because it is so cold. The average temperature on the surface of Mars is -63°C (-81°F), and there appears to be no vegetation on Mars.

1. What makes Earth habitable for all life forms that we know of?

    A) Earth's distance from the sun
    B) the existence of liquid water on Earth
    C) Earth's unique atmosphere
    D) all of the above

Simple Solutions® Science   Level 5

For each of the planets listed below, list two characteristics that make it impossible for the planet to support life as we know it.

2. Mercury _____ _____

3. Venus _____ _____

4. Mars _____

5. A _____ is a combination of two or more substances.

6. Which one of these will always take up the same amount of space and will keep its shape no matter what container it is in?

   solid   liquid   gas   all three

7. Dead, decaying plants and animals are broken down by decomposers. During this chemical change, what is happening to the atoms that make up the decaying matter?

   A) The atoms do not change.
   B) The atoms are changing in form.
   C) The atoms are being destroyed.
   D) New atoms are being created.

8. When the number of electrons is greater than the number of protons in an atom, the atom is _____.

   neutral   unstable   positively charged   negatively charged

9. The Northern Hemisphere has winter while the Southern Hemisphere has _____.

   winter   spring   summer   fall

10. An object at rest will _____ if no unbalanced force is applied.

    increase in velocity   remain at rest   decrease in speed   accelerate

Lesson #131

The Outer Planets

A large asteroid belt separates the inner planets from the outer planets, and the outer planets are the farthest from the sun. These planets are Jupiter, Saturn, Uranus, and Neptune. The outer planets are similar to each other, but they have their differences as well. The outer planets are nicknamed the "Gas Giants" because they are composed mostly of hydrogen and helium. (Remember, the inner planets are mostly made of dense rock.) And the outer planets are called "giants" because they are all very large compared to the inner planets.

Jupiter is the fifth planet from the sun, and it is the largest planet in the solar system. Jupiter has a giant "red spot" which is a storm that has been going on for hundreds of years. More than sixty moons revolve around the planet Jupiter. The next planet from the sun is **Saturn**, and it is well known for the remarkable rings that orbit the planet. These rings are composed of ice, rock, and frozen gases. Saturn also has about sixty moons. **Uranus** comes next; it is a pale blue ball of gas with many rings and over 25 moons. Uranus orbits the sun tilted on its side. The farthest of the outer planets is **Neptune**, a planet that is deep blue in color. Neptune is the windiest planet that we know of, with winds up to 1,243 miles per hour!

In addition to the eight planets, the solar system also contains several dwarf planets: Pluto, Ceres, Eris, Haumea, and Makemake. A **dwarf planet** is a round body that orbits the sun. However, dwarf planets are much smaller than regular planets; the four that we know of so far are smaller than Earth's moon. Also, dwarf planets have not cleared the neighborhood around their orbit. That means the dwarf planets have an orbit that is not clear; they may mingle with asteroid belts or move in and out of the orbits of other planets.

1. What is one **distinction** between the inner planets and the outer planets?

   _____

2. What is one **distinction** between a regular planet and a dwarf planet?

   A) A regular planet is round; a dwarf planet is not.
   B) A dwarf planet does not revolve or rotate around the sun.
   C) A regular planet is made of rock; a dwarf planet is made of gases.
   D) A dwarf planet is smaller and does not have a clear orbit.

3. Which of these have an orbit? Underline any that do.

   sun   moon   Earth   Pluto   Mars

4. What is an **orbit**?

    A) a path that a body in space follows around another body in space
    B) a force that pulls an object in space toward its center
    C) a system for telling time or predicting weather patterns
    D) a body that is spherical in shape and is made of dense rock

5. The Northern Hemisphere is in which season when it is fall in the Southern Hemisphere?

    winter   spring   summer   fall

6. Which scientific instrument would you use to measure temperature?

    _____

7. The more thermal energy a body of mass has the more _____ it has.

    pollution   kinetic energy   volume   air pressure

8. What are two of the characteristics of Earth that make it able to support life as we know it?

    A) Earth's shape and the fact that there are active volcanoes
    B) the landscape of Earth and its moon
    C) Earth's unique atmosphere and distance from the sun
    D) wind, gases, and the rock cycle

9 – 10. Look at the illustration below. Explain how simple machines are helping to decrease the amount of work needed to move a load.

Man pushing a wheelbarrow up an inclined plank

Lesson #132

1. Another word for an inclined plane is a _____.

2. One of these has neither a definite shape, nor a definite volume; it will take up as much space as there is. Which is it?

   solid   liquid   gas

3. Name an *abiotic* factor that can cause change in an environment.

   _____

4. This is a drawing of a **complete circuit**. What can be expected to happen to both light bulbs in this series?

   A) They will both light up.
   B) Only one will light up at a time.
   C) Neither bulb will light since there is no switch.
   D) The bulbs will probably blink on and off.

5. Which of these is not an example of a chemical reaction?

   A) fireworks   C) burning wood
   B) boiling water   D) digestion

6. Put these steps of the scientific method in order. Number from 1 – 4.

   _____ Develop a hypothesis that may answer your question or solve your problem.

   _____ Conduct an experiment; record data.

   _____ Form a conclusion based on the results of your experiment.

   _____ Start with a problem or question.

7. Write each word next to the hint that describes it.

omnivore   primary consumer   producer   food chain
decomposer   tertiary consumer   secondary consumer

A) breaks down organic matter   _____

B) eats animals that eat plants   _____

C) eats both plants and animals   _____

D) beginning of the food chain   _____

E) eats plants only   _____

F) eats animals that eat other animals _____

8 – 10. Using words and/or drawings explain why Newton's first law of motion is a good reason why people riding in cars should wear safety belts.

Lesson #133

Scientific Inquiry: Crossword Review

Use the words from the word bank and the clues below to complete the crossword puzzle.

conclusion	conclusive	constant	question	control
evidence	experiment	hypothesis	verify	data
replicate	research	variable	inconclusive	

Across

2. information gathered from books, magazines, the internet, and other people
6. procedure designed to test a hypothesis
8. information gathered during an experiment
10. a problem to solve; the first step of the scientific method
12. uncertain; open to doubt
13. prove
14. a factor that does not change in an experiment

Down

1. any factor that can change in an experiment
3. an educated guess; a part of the scientific method
4. repeat
5. definite; certain
7. the group that does not receive the experimental treatment
9. final step of the scientific method; report or demonstration of data gathered during an experiment
11. data, facts, and information that may support a hypothesis

Simple Solutions© Science
Level 5

Lesson #134

1. Water turns to vapor (gas) in a process called _____.

    melting   freezing   condensation   evaporation

2. Terrance poured himself a glass of soda and left it on his desk overnight. The next day, the soda was flat; there was no carbonation (fizz) left in the soda. What kind of change took place in the glass of soda?

    physical   chemical   both   neither

3. Which of these items in Cindy's desk drawer would be good conductors of electricity?

     erasers
     steel paper clips
     plastic ruler
     staples

4. About how long does it take for the moon to go through its phases?

    24 hours   7 days   21 days   30 days   365 days

5. What makes a substance a **mixture**?

    A) It's a combination of two or more substances.
    B) It has been heated to high temperature.
    C) The combination makes something new.
    D) It has caused a chemical reaction.

6. Organisms that make their own food in a process called photosynthesis are called _____.

    producers   consumers   decomposers   food webs

7. An atom will have a positive charge when _____
    A) it has more electrons than protons.
    B) it has more protons than electrons.
    C) it has no neutrons.
    D) it has the same number of protons, neutrons, and electrons.

8. One of these will always take up the same amount of space, but it will take on the shape of the container that it is in. Which is it?

    gas   liquid   solid   all three

9. One of these organisms is not like the others. Which one doesn't belong?

    whale   jellyfish   squirrel   salamander

10. Explain why the one you chose doesn't belong with the other three.

    _____

    _____

Lesson #135

1. Jason's mother is a chemist. She says that you can burn metal called magnesium in an open-air container. And when the metal is fully burned, a different substance called magnesium oxide is formed. Jason's mom says that after completely burning, the magnesium oxide has a mass that is greater than the magnesium that was burned. How can this be?

   A) The magnesium reacts with oxygen in the air during the burning and gains oxygen atoms.
   B) Burning creates more mass.
   C) Burning is a chemical reaction which destroys mass.
   D) Jason's mother doesn't understand the Law of Conservation of Matter.

2. What is the name for materials that <u>do not</u> allow electricity to flow freely through them?

   A) conductors
   B) buzzers or switches
   C) metals or graphite
   D) insulators or non-conductors

3. What is the purpose of a battery in a series circuit?

   A) It provides electricity.
   B) It provides an insulator.
   C) It is a type of switch.
   D) It uses electricity.

4. Why do the cups and saucers in the photo remain stacked?

   A) Balanced forces are acting upon them.
   B) Unbalanced forces are acting upon them.
   C) No forces are acting upon them at this time.
   D) Buoyant force is holding them in place.

5. What is the job of the nucleus of a cell?

   A) to fill in the space inside the cell
   B) hold the cell in place
   C) to control the cell's activities
   D) none of these

6. The troposphere contains _____ of the gases in the atmosphere.

         25%   90%   10%   100%

7. Which season will the Southern Hemisphere be experiencing when it is spring in the Northern Hemisphere?

       winter   spring   summer   fall

8. What do these terms have in common?

   nitrogen   water vapor   oxygen   dust particles   carbon dioxide

   A) These are all names of gases.
   B) They all list types of solutions.
   C) They name particles that make up the atmosphere.
   D) They are all parts of the rock cycle.

9. Which <u>two</u> of these words may describe a tertiary or third-level consumer?

       carnivore   producer   omnivore   herbivore

10. What will happen to a circuit that is **incomplete**?

    A) The circuit will work.
    B) The circuit will not work.
    C) The circuit may or may not work
    D) The circuit will work for a short time.

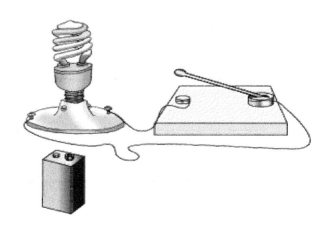

Lesson #136

1. Which two organisms belong to the same kingdom?

   aloe plant   diatom   amoeba   centipede

2. Why are the items in these photos called "simple" machines?

   A) They are not very useful.
   B) They were invented a long time ago.
   C) They have no moving parts.
   D) They are easy to use.

3. What are the three states of matter?

   _____ _____ _____

4. Decomposers break down organic matter and _____.

   A) put nutrients back into an ecosystem
   B) make their own food through photosynthesis
   C) both A and B
   D) none of the above

5. What is Newton's first law of motion, and how does this cartoon help to illustrate the concept?

   _____

   _____

   _____

   _____

   _____

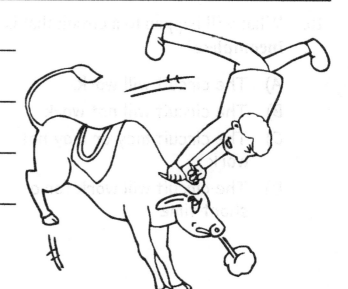

6. Fill in the chart with the missing seasons and hemispheres.

When it is ___	in the ___ Hemisphere,	it is ___	in the ___ Hemisphere.
winter	Northern	A)	B)
C)	Southern	summer	D)
E)	F)	spring	Southern
G)	Southern	fall	H)

7. Look at the photo. Which of these is a scientific term that would explain why the cans are not falling over?

    magic   inertia   humidity   atmospheric pressure

8. What would cause the cans in the photo to topple over?

    A) a balanced force   C) any force
    B) an unbalanced force   D) no force

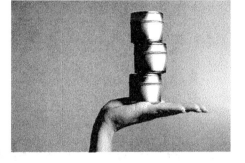

9. Three of these organisms have something in common. Which one doesn't belong?

    squirrel   turkey vulture   groundhog   deer

10. Explain why the one you chose doesn't belong with the other three.

    _____

    _____

Lesson #137

1. A faucet handle can use at least two simple machines. What are they?

   A) _____

   B) _____

2. Burning a marshmallow is an example of a <u>chemical</u> change because ___.
   A) carbon is released
   B) during burning molecules are rearranged
   C) sugar molecules in the marshmallow break down into carbon and water
   D) all of these

3. Why are the outer planets called the "gas giants?"

   _____

   _____

4. The diagram shows the orbits of the _____.

   inner planets   outer planets
   moons   dwarf planets

5. When the number of protons is greater than the number of electrons in an atom, the atom is _____.

   neutral   unstable
   positively charged   negatively charged

6. At which of these do temperatures tend to be the warmest?

   north pole   south pole   equator   none of these

7. Which consumer eats only plants?

   herbivore   carnivore   omnivore   none of these

Simple Solutions© Science   Level 5

8. The air pressure in the atmosphere is greater _____.

   A) near the moon
   B) the higher up you go
   C) the closer you are to Earth's surface
   D) at the top of a mountain

9. True or False?

   A) _____ A force is needed in order to change motion.

   B) _____ Acceleration is a measure of the change in velocity.

   C) _____ Velocity is a measure of speed in a certain direction.

10. Betty is baking a birthday cake, and she is wondering whether what she is doing is causing physical or chemical changes. Read each phrase and decide which it is. Write **P** for physical change, **C** for chemical change, or **N** for neither.

    A) _____ cracking open two eggs

    B) _____ sifting together flour, sugar, salt, and baking soda

    C) _____ blending all ingredients with an electric mixer

    D) _____ baking the cake for 40 minutes at 350°F

    E) _____ melting butter for the icing

    F) _____ spreading icing on the cake and putting candles on top

    G) _____ lighting the candles

    H) _____ singing "Happy Birthday"

275

Lesson #138

1. Which is true?

    _____ All solutions are mixtures.

    _____ All mixtures are solutions.

    _____ All solutions and mixtures contain water.

2. Which of these is not an example of a physical change?

    A) popping popcorn
    B) rusting metal
    C) coloring eggshells
    D) evaporating water

3. Which of the following are *abiotic* factors that can cause change in an environment?

    earthquake   cell division   weathering   disease

4. Fill in the blanks with **solid**, **liquid**, or **gas**.

    A) A _____ keeps its shape and volume.

    B) A _____ keeps its volume but changes its shape according to its container.

    C) A _____ does not keep its shape or its volume. It expands to take up all the space there is.

5. Natural resources that can be used over and over again – if used wisely – are called _____.

    renewable resources   nonrenewable resources   both of these

6. Sara pushes and the car begins to move. Why?

    A) There are no forces acting on the car.
    B) The force put forth by the car is equal to the force put forth by Sara.
    C) Forces acting on the car are balanced.
    D) Forces acting on the car are unbalanced.

7. Electricity does not move easily through wood and plastic; these materials are called _____.

   magnetic   insulators   conductors   none of these

8. Plants get nutrients, light, and water from the environment. They make food for themselves and other organisms, and they put oxygen back into the air. What is this process called?

   _____

9. What is one distinction between Earth and Venus?

   A) Venus revolves around the sun; Earth does not.
   B) Earth is mostly made of rock; Venus is mostly gaseous.
   C) Venus is Earth's "twin."
   D) Earth has an atmosphere that makes it habitable; Venus does not.

10. **Name the Category** Study each list of words and decide what all the words have in common. Then write a category name at the top of each list.

A)	B)	C)	D)	E)
microscope	nitrogen	squirrel	coal	screw
thermometer	oxygen	blue whale	oil	wedge
barometer	carbon dioxide	dog	natural gas	pulley
hygrometer	water vapor	bat	old-growth forests	lever

Lesson #139

1. About how long does it take for the moon to go through its phases?

   48 hours   21 days   365 ¼ days   30 days   100 days

2. What is one distinction between a planet and a dwarf planet?

   A) A planet has cleared the neighborhood around its orbit, but a dwarf planet has not.
   B) A dwarf planet is much smaller than a regular planet.
   C) A planet stays within its own orbit, but a dwarf planet may mingle in the orbit of an asteroid belt or cross into the orbits of other planets.
   D) All of the above

3. During photosynthesis, plants use sunlight, water, and carbon dioxide to create glucose, a form of sugar. During this chemical change, _____

   A) new matter is created.
   B) old matter is destroyed.
   C) molecules are rearranged to form a new substance.
   D) all of the above occur.

4. _____ is the amount of water vapor in the air compared to the maximum amount possible.

   Relative humidity   Air pressure   Pollution index   Standard time

5. Which organism completely changes its physical features and food source when it becomes an adult?

rabbit

tadpole

mouse

turtle

Simple Solutions© Science Level 5

6. What is the Law of Conservation of Matter?

   _____

   _____

7. Which simple machine allows this cart to move easily across the floor?

   _____

8. A) Name the four inner planets.

   _____

   _____

   _____

   _____

   B) Earth is the _____ planet from the sun.

       first   second   third   fourth

9. Which organism belongs to the same kingdom as birds?

     apple tree   bacteria   mushroom   none of these

10. Jerry put together a simple circuit that looks like the one pictured below. Why doesn't the bulb in Jerry's circuit light up?

    A) The battery is probably dead.
    B) The bulb is burned out.
    C) The circuit is not complete.
    D) The wire is not insulated.

279

Lesson #140

1. A _____ is a simple machine that has a slanted side and a sharp edge for sliding or for cutting.

2. How long does it take for Earth to complete one full rotation?
    30 days   365 days   24 hours   60 minutes

3. Which materials listed below are good conductors of electricity?
    silver   rubber   plastic   copper   chalk

4. Which materials listed below are good insulators?
    foam   steel   rubber   tap water   glass

5. Old cars, nails, and other things made of metal get rusty in a chemical reaction called corrosion. Oxygen reacts with the metal and the result is rust. A rusty nail has more mass than it had when it was new. What is the best explanation for this?

    A) The rusty nail has its original mass plus oxygen from the air.
    B) The corrosion process has created mass.
    C) Organisms have begun to grow on the nail.
    D) The nail is holding water which increases its weight.

6. When vapor (gas) changes to water (liquid), the process is called _____.

    condensation   evaporation   melting   freezing

7 – 10. Sort the terms listed below into five different categories. Give each category a title. Then write your titles in the top spaces and list the terms under each title.

question	protists	research	air	omnivore
consumer	producer	shelter	data	cell membrane
herbivore	carnivore	bacteria	fungi	conclusion
cytoplasm	water	vacuoles	plant	mitochondria
experiment	animal	nucleus	food	waste disposal

Category #1	Category #2	Category #3	Category #4	Category #5

Level 5

Science

Help Pages

Help Pages

Glossary

Abiotic	not biotic; refers to things that are not and never were alive (rocks, minerals, soil, water, sunlight, and air) (Lesson #24)
Absolute Zero	the coldest point possible; temperature at which all motion of particles stops (Lesson #84)
Acceleration	the rate at which velocity changes; acceleration may increase or decrease (Lesson #122)
Air Pressure	the weight of the atmosphere pressing toward Earth's core (Lesson #49)
Alto	from the Latin, altus, meaning "high." When used to name a cloud, it identifies clouds that look high, but are really middle clouds (Lesson #59)
Amphibian	a cold-blooded vertebrate that hatches in water and is born with gills but develops lungs and lives on land as an adult (Lesson #19)
Anemometer	instrument that measure wind speed (Lesson #54)
Appliance	part of a circuit; anything that is powered by electric current (Lesson #101)
Atmosphere	a blanket of air surrounding the Earth and made up of several layers: troposphere, stratosphere, mesosphere, thermosphere, and exosphere (Lesson #49)
Atom	is the smallest bit of any given type of matter (Lesson #94)
Balance	tool used to measure the mass of objects (chart)
Balanced Forces	forces that are equal in strength and cancel each other out (Lesson #114)
Barometer	weather instrument used to measure air pressure (Lesson #54)
Bedrock	layer of soil beneath the subsoil
Biomass	organic material that can be burned to release energy (Lesson #91)
Bird	one of five vertebrate groups; all birds have beaks, wings, and bodies that are covered with feathers (chart)
Body System	a group of organs working together within the body (nervous system, circulatory system, digestive system, etc.) (Lesson #16)
Botanist	a scientist who specializes in the study of plants (Lesson #21)
Buoyancy	the force of water pushing up on an object and keeping it afloat (Lesson #115)
Carbon Dioxide	one of the gases found in Earth's atmosphere; plants take in carbon dioxide during photosynthesis

Simple Solutions© Science Level 5

Help Pages

Glossary

Carnivore	meat-eating; an animal that eats other animals (Lesson #12)
Carrying Capacity	population size that a certain ecosystem can support without damaging the ecosystem (Lesson #18)
Cell Division	the way cells make more cells; the nucleus divides itself, and one cell becomes two cells (Lesson #16)
Cell Membrane	a protective covering that allows nutrients to be absorbed into a cell and allows waste to pass out of the cell (Lesson #13)
Cell Wall	an outer covering of plant cells which helps the plant cells stick together and gives support to the plant (Lesson #14)
Celsius	one of three types of thermometers used to measure temperature or average kinetic energy (Lesson #84)
Cementation	to bond pieces together like cement; one of the processes by which sedimentary rock is formed
Characteristics	properties; anything that describes a substance
Chemical Change/Reaction	a change in matter in which an entirely new substance is formed; atoms are rearranged during a chemical reaction (Lesson #74)
Chemical Potential Energy	stored energy that can be converted to kinetic energy (Lesson #79)
Chlorophyll	the substance in plant cells that causes the green color in plants and enables plants to make food through photosynthesis (Lesson #7)
Chloroplast	cell part that makes food for the plant; found only in plant cells (Lesson #14)
Chrysalis	cocoon; pupa; the third stage of complete metamorphosis of insects (Lesson #21)
Circuit	a path that electricity follows (Lesson #101)
Cirrus	Latin for "curl of hair;" wispy, featherlike, high altitude clouds (Lesson #59)
Classification	a system that scientists use to organize living and non-living things (Lesson #15)
Clay	a very fine-grained soil which does not easily allow air and water to pass through it (Lesson #32)
Climate	established weather pattern of a certain area over a long period of time (Lesson #58)
Closed Circuit	a circuit with no breaks or open switches; allows electricity to freely flow through it (Lesson #102)

Help Pages

Glossary

Cold Front	condition that occurs when cold air moves in to replace warmer air; may bring heavy rain, thunderstorms, hail, or snow (Lesson #57)
Community	all of the populations that live in an ecosystem at one time
Compaction	pressing together with great force/pressure; one of the processes by which sedimentary rock is formed
Complete Circuit	a closed circuit (Lesson #102)
Conclusion	final step of the scientific method; report of results of an inquiry (Lesson #6)
Conclusive	definite; certain; results that are not questionable (Lesson #14)
Condensation	process by which water vapor turns to liquid water (Lesson #40)
Condensation Point	the temperature at which a given substance changes from gas to liquid; the same temperature as the boiling point for that material (Lesson #67)
Conduction	a transfer of heat that occurs when a heat source comes into contact with other mass that is cooler (Lesson #86)
Conductor	substance which allows electricity to flow easily through it (metals, dried wood, tap water) (Lesson #99)
Conifer	type of plant that produces its seeds within a cone
Constant	unchanging; in an experiment, a variable that does not change (Lesson #9)
Consumer	any organism that gets energy by consuming (eating) other organisms (Lesson #10)
Control Group	the persons or items in an experiment that do not receive the experimental treatment (Lesson #11)
Convection	transfer of heat through the circulation of a gas or liquid as it warms and then cools (Lesson #87)
Corrosion	rust; a chemical reaction in which water and oxygen react with metal
Crust	Earth's surface; the outermost layer of Earth (Lesson #28 & chart)
Crystalline Solid	formed with atoms arranged in a definite repeating pattern; a crystal (Lesson #37)
Cumulus	Latin for "heap" or "pile"; dense, white, fluffy clouds (Lesson #59)

Help Pages

Glossary

Current Electricity	electricity that moves through wires (Lesson #99)
Cytoplasm	a jelly-like substance that fills cells (Lesson #13)
Data	facts, statistics and other information (Lesson #5)
Deciduous	plants that shed their leaves at the end of the growing season
Decomposer	organisms that break down the remains of other organisms and return vital nutrients to the soil; bacteria, protists, fungi, earthworms, etc. (Lesson #10)
Deforestation	the permanent destruction of forests caused by the cutting of too many trees too quickly (Lesson #37)
Demonstration	a way of showing a scientific concept (Lesson #38)
Density	a measure of how closely molecules are packed in a given amount of space (Lesson #69)
Dependent Variable	a variable that changes, depending upon other factors in the experiment; a dependent variable may change.
Desert	an ecosystem which is very dry, getting less than 10 inches of rain per year
Dwarf Planet	is a round body that orbits the sun but is much smaller than a regular planet and has not cleared the neighborhood around its orbit (Lesson #131)
Ecosystem	all of the living and nonliving things interacting and affecting each other in a certain area (Lesson #18)
Elastic Potential Energy	the type of stored energy that is in a stretched-out rubber band or spring (Lesson #79)
Electricity	energy produced by the movement of electrons (Lesson #97)
Electron	negatively charged particle that spins in an orbit around the nucleus of an atom (Lesson #95)
Element	a substance made up of only one kind of atom (Lesson #94)
Elliptical	egg-shaped; the shape of the orbit of anything that travels around the sun in the solar system (Lesson #127)
Energy	the ability to do work (Lesson #77)
Energy Conservation	a way to protect and preserve natural resources by recycling, using less, and reusing materials (Lesson #92)
Energy Efficiency	using technology to reduce the amount of waste when using natural resources (Lesson #92)

Help Pages

Glossary

Entomology	the study of insects (Lesson #21)
Environment	the natural world; everything around us
Environmentalist	scientist who studies the natural world and those who work to protect it (Lesson #93)
Erosion	process by which broken down rocks are carried away by wind, water, or moving ice (Lesson #27)
Evaporation	process by which a liquid changes to vapor or gas (Lesson #40)
Evidence	facts that support conclusions (Lesson #14)
Experiment	one of the steps of the scientific method; a set of procedures meant to test a hypothesis (Lesson #4)
Fahrenheit	one of three temperature scales; used to measure temperature or average kinetic energy (Lesson #84)
Fish	one of the five vertebrate groups; all fish live in water, breathe through gills, and are covered with scales (chart)
Food Chain	process by which energy travels between organisms (Lesson #5)
Force	a push or a pull (Lesson #114)
Fossil	the imprint or remains of things that lived long ago (Lesson #29)
Fossil Fuels	nonrenewable resources formed from the remains of organisms over the past several hundred years; coal, oil, natural gas (Lesson #89)
Frame Of Reference	everything that is not moving around an object that is in motion (Lesson #119)
Freezing Point	temperature at which a liquid will become a solid (Lesson #67)
Frequency	number of vibrations per second (Lesson #107)
Freshwater	water that does not have a large amount of salt in it (lakes, streams, rivers, ponds)
Friction	a force created by two objects rubbing against each other; a force that reduces motion by working against it (Lesson #115)
Fulcrum	the fixed point on a lever (chart)
Gas	one of the three states of matter; substance made of widely-spaced particles that break away from each other easily (Lesson #65)
Geothermal Energy	heat energy that comes from the Earth (Lesson #83)

Help Pages

Glossary

Grassland	a somewhat dry, flat ecosystem where the main vegetation is grass
Gravitational Potential Energy	energy created by an attraction between two objects; gravity pulls objects toward Earth (Lesson #79)
Gravity	a force that pulls objects toward each other
Ground Water	water that has soaked into the ground and has collected in underground reservoirs (Lesson #41)
Habitable	able to support life (Lesson #130)
Habitat	the place where an organism lives
Hand Lens	a hand-held magnifying glass
Hardness	one of the physical properties of minerals (Lesson #37)
Hazardous Waste	harmful pollutants that contaminate the environment
Heat Energy	thermal energy created by the movement of atoms; the more heat energy an object absorbs, the more its kinetic energy will increase (Lessons #83 - 84)
Herbivore	an animal that eats only plants (Lesson #12)
Hibernation	a very inactive state in a safe, hidden place; animals hibernate when temperatures are cold and food is scarce (winter)
Homogeneous	all the same
Humidity	a measure of the amount of moisture in the air (Lessons #54 - 55)
Humus	the decayed remains of plants and animals (Lesson #32)
Hydropower	water power (Lesson #91)
Hydrosphere	all the waters of the Earth (Lesson #36)
Hygrometer	a weather instrument that measures the humidity level or how much moisture is in the air (Lesson #54 - 55)
Hypothesis	an educated guess (Lesson #3)
Igneous	one of the three kinds of rock; it is formed by the cooling and hardening of molten rock (Lesson #31)
Immunization	medicine meant to prevent a certain type of disease (Lesson #44)
Inclined Plane	a slanted surface used to move things to higher or lower places; one of the simple machines (chart)

Help Pages

Glossary

Incomplete Metamorphosis	growing through three different life stages: egg, nymph, and adult; except for size, the insect looks mostly the same during the second two phases (chart)
Inconclusive	not proving anything; results of an experiment are inconclusive if they neither prove nor disprove the hypothesis (Lesson #4)
Independent Variable	the variable (in an experiment) that the experimenter will change to see what effect it has on the dependent variable (Lesson #9)
Inertia	the tendency of an object not to move unless a force acts upon it; the tendency for a moving object to continue moving unless a force acts upon it (Lesson #118)
Inhabitable	habitable; able to support life (Lesson #130)
Inner Core	the solid sphere-shaped center of the Earth; the core is made of iron and nickel (Lesson #28 & chart)
Inner Planets	the four planets closest to the sun: Mercury, Venus, Earth, and Mars (Lesson #130)
Inorganic	not related to living organisms; example: minerals (Lesson #30)
Instinct	natural impulse; behavior an animal knows without being taught
Insulator	non-conductor; material that is a poor conductor of electricity (rubber, glass, paper) (Lesson #100)
Invertebrates	multi-celled organisms that do not have backbones (chart)
Investigation	inquiry (Lesson #1)
Kelvin	one of three temperature scales used to measure temperature or average kinetic energy (Lesson #84)
Kinetic Energy	energy in motion (Lessons #78 – 79)
Kingdom	one of the major groupings that scientists use to organize and classify living things (Lesson #17)
Larva	the second of the four stages of complete metamorphosis; a larva hatches from an egg (chart)
Law of Conservation of Matter	law that states that matter is neither created nor destroyed (Lesson #75)
Lever	a type of simple machine; a bar that pivots on a fulcrum to move or lift heavy loads (chart)
Light Energy	a type of energy that travels in waves (Lesson #109)

Simple Solutions© Science — Level 5

Help Pages

Glossary

Liquid	one of the three states of matter; liquids have a definite volume but take on the shape of their containers (Lesson #64)
Lithosphere	Earth's crust and the top part of the mantle
Loam	a mixture of soil that contains sand, silt, and clay along with humus, water, and air; the best soil for growing plants (Lesson #32)
Lunar Cycle	the pattern of the phases of the moon (new, waxing crescent, first quarter, waxing gibbous, waning gibbous, last quarter, waning crescent, full) that takes about 30 days to cycle through
Luster	one of the physical properties that scientists use to classify minerals: the way light reflects off of the mineral (Lesson #37)
Magma	melted rock (Lesson #33)
Magnet	a substance that is magnetic; a substance that attracts iron, nickel, or cobalt (Lesson #105)
Mammal	a warm-blooded animal whose body is covered with hair or fur, is able to regulate its body temperature, and feeds its young with mother's milk (chart)
Mantle	the thickest layer of Earth located between crust and outer core (Lesson #28 & chart)
Mass	the amount of matter in an object; mass can be measured using a balance (Lesson #62)
Materials List	items needed to complete an experiment (Lesson #4)
Matter	anything that has volume and mass (Lesson #61)
Melting Point	the temperature at which a solid will become a liquid (Lesson #67)
Mesosphere	one of the outer layers of the atmosphere; the mesosphere is beyond the troposphere and the stratosphere (Lesson #50)
Metamorphic	one of the three types of rock; forms when high heat and pressure change a rock's shape and substance into a new type of rock (Lesson #31)
Metamorphosis	changing shape; the series of changes in phase and appearance from birth to adulthood (Lesson #21 & chart)
Meteorology	the study of the atmosphere and weather conditions (Lesson #54)
Microscope	a lab instrument that magnifies the view of tiny objects to hundreds of times their natural size (chart)

Help Pages

Glossary

Migration	an instinctual animal behavior; the seasonal movement of animals to places that are warmer, safer, or have a better food supply
Mimicry	imitating the look of another animal; an instinctual self-defense behavior
Mineral	a naturally occurring inorganic solid (Lesson #30)
Mixture	a combination of two or more substances (Lesson #70)
Model	a replica or smaller version meant to show the characteristics of something (model airplane, human heart, erosion, etc.) (Lesson #38)
Molecule	the atoms of two or more different elements joined together (Lesson #94)
Moon Phases	appearance of the moon at different times during a thirty-day cycle; caused by the sun's shadow blocking the light that is reflected off of the moon
Motion	movement; can only be stopped or started by a force acting on an object (Lessons #114 & 118)
Natural Resources	all of the naturally occuring materials that humans and other organisms use for survival (air, water, trees, coal, oil, animals, etc.) (Lesson #89)
Neutron	atomic particle that has a neutral (neither positive nor negative) charge (Lesson #95)
Nimbus	Latin word for "cloud" that always signals rain (Lesson #59)
Non-Conductor	insulator; material that is a poor conductor of electricity (rubber, glass, paper) (Lesson #100)
Nonliving	not alive
Nonrenewable Resources	natural resources that cannot be replaced within a person's lifetime (Lesson #89)
Nourishment	food, water, and other nutrients; anything that feeds an organism
Nucleus	the part of a cell that controls the cell's activities; the center of an atom which contains protons and neutrons (Lessons #13)
Nutrients	anything that nourishes or feeds an organism (food, water, vitamins, minerals, etc.)
Omnivore	an animal that eats both plants and other animals (Lesson #12)
Opaque	material that does not allow light to pass through it (Lesson #37)

Simple Solutions© Science　　Level 5

Help Pages

Glossary

Open Circuit	an incomplete circuit; a circuit which does not allow for the complete flow of an electrical current through it (Lesson #102)
Orbit	a path that a body in space follows around another body in space (Lesson #127)
Organic Material	any substance that comes from living or once living things
Organism	a living thing (Lesson #5)
Outer Core	the liquid layer of molten rock nearest to the hard center of the Earth (Lesson #28 & chart)
Outer Planets	the four planets beyond the asteroid belt and farthest from the sun: Jupiter, Saturn, Uranus, and Neptune (Lesson #131)
Ozone Layer	the layer of atmosphere that protects life on Earth by absorbing harmful ultraviolet radiation from the sun (Lesson #49)
Parallel Circuit	a circuit with multiple paths which are side by side and receive electric current from the same source, but carry the current to separate receivers (Lesson #103)
Periodic Table of Elements	an organized list of all known elements (Lesson #94)
Photosynthesis	process by which green plants make their own food using sunlight, water, and carbon dioxide (Lessons #1, #3, & #4)
Physical Change	a change in the phase or state of matter which does not change what the substance is at a molecular level (Lesson #73)
Physical State or Phase	one of the properties of matter (solid, liquid, and gas are 3 states of matter) (Lesson #61)
Physics	the study of matter and energy (Lesson #93)
Pitch	how high or low a sound is (Lesson #107)
Planet	a large mass that has settled into a nearly spherical (round) shape, orbits a star, and has cleared the neighborhood around its orbit (Lesson #130)
Population	a certain group of the same kind of organism living in an ecosystem (Lesson #18)
Potential Energy	stored-up energy (Lesson #78)
Precipitation	rain, snow, sleet, hail, fog, dew or water in any form that falls to Earth's surface (Lesson #39)
Predator	an animal that hunts another animal as food

Help Pages

Glossary

Prevailing Winds	the constant flow of air created by the movement of cooler air into warmer areas (Lesson #53)
Prey	an animal that is hunted by another animal as food
Primary Consumer	herbivores; animals that only eat producers (plants) (Lesson #12)
Procedure	a step by step list of what to do while performing an experiment (Lesson #4)
Process	an on-going movement or series of changes (weathering, rock cycle, water cycle, photosynthesis, cementation, freezing, etc.)
Producer	an organism, such as a green plant that can make its own food (Lesson #8)
Properties	characteristics; anything that describes a substance
Proton	atomic particle that has a positive charge (Lesson #95)
Pulley	a simple machine that uses grooved wheels and ropes to raise and lower objects (chart)
Pupa	the third of the four stages of complete metamorphosis; the pupa is also called a cocoon or chrysalis (chart)
Question	the first step of the scientific method; what the investigation or inquiry is meant to answer (Lesson #1)
Radiant Energy	energy that moves in waves (light waves, radio waves, microwaves, x-rays) (Lesson #79)
Radiation	one of the three ways that thermal energy is transferred; movement of heat energy through waves (Lesson #87)
Recycle	use again; to save resources and the environment by reusing materials instead of disposing of them (Lesson #92)
Reflection	light bouncing off a shiny or smooth surface; creates a mirror image (Lesson #109)
Refraction	the bending of light as it passes from one substance to another, such as from air to water (Lesson #109)
Renewable Resources	natural resources that are replaced by natural ecological cycles and when used wisely, can be used over and over again (Lesson #89)
Replicate	repeat (Lesson #14)
Reptile	one of the vertebrate groups; animal whose body is covered with scales and breathes through lungs (chart)

Help Pages

Glossary

Research	the second step of the scientific method; to investigate and explore in order to find more information about a topic of inquiry (Lesson #2)
Rock	a natural substance made of one or more minerals
Rock Cycle	process by which rocks constantly change from one form to another (igneous, sedimentary, metamorphic) (Lesson #31)
Root	the part of a plant that anchors the plant in the soil and takes in water and nutrients from the soil
Runoff	melting ice or snow and precipitation that drains off the land and soaks into the ground or flows toward a body of water (Lesson #41)
Sand	a type of soil that has very large particles, is loose, and feels coarse or rough (Lesson #32)
Scavenger	an animal that feeds on the remains of dead animals and helps to clean up the environment by getting rid of decaying organic matter (Lesson #12)
Scientific Inquiry	an organized way to find answers to questions or solutions to problems (Lesson #1)
Scientific Method	a series of steps that includes asking a question, doing research, formulating a hypothesis, experimenting, gathering data, and drawing conclusions (Lesson #1)
Scratch Test	test to determine the hardness of a mineral; a harder mineral will scratch a softer mineral (Lesson #37)
Screw	a type of simple machine; an inclined plane spiraled around a post used to fasten or hold things together (chart)
Secondary Consumer	a carnivore; an animal that eats other animals (Lesson #12)
Sedimentary	a type of rock formed when sediments bond together by pressure over time (Lesson #31)
Seed	the first stage of life for many plants; contains the food to help a new plant grow
Series Circuit	an electrical circuit that has only one pathway from the source, through the conductor, to the receiver (Lesson #103)
Silt	smooth, powdery soil made of small particles (Lesson #32)
Simple Machines	machines that have only a few or no moving parts and need a single force such as a push, a pull, or a lift to make them work (chart)

Help Pages

Glossary

Soil	a mixture of broken down rock, air, water, and organic material (Lesson #32)
Solar Power	energy that comes from the sun (Lesson #81)
Solid	one of the three states of matter; solids have a definite shape and volume (Lesson #63)
Solidification	moving from a liquid or gas to a solid state; freezing (Lesson #66)
Solution	a type of mixture in which all the parts are evenly distributed (Lesson #71)
Sound	a type of energy that is created by vibrations and travels in waves (Lesson #107)
Source (Power Source)	the supply of power in an electric circuit; may be a battery, generator, or electrical outlet (Lesson #101)
Spring Scale	a tool used to measure weight or friction (chart)
States of Matter	solid, liquid, and gas (Lesson #61)
Static Electricity	a type of potential energy that builds up on an object as the result of freed electrons (Lesson #98)
Stationary Front	a barely moving mass of air (Lesson #57)
Stem	the part of a plant that grows above ground; gives the plant support and carries water and nutrients from the roots to the rest of the plant
Stratosphere	layer of Earth's atmosphere that is closest to the troposphere and contains most of the ozone (Lesson #50)
Stratus	Latin for "spread out"; clouds that are layered and look like blankets or mattresses (Lesson #59)
Streak	one of the properties of minerals; the color of the mark that a mineral leaves when it is slid over a streak plate (Lesson #37)
Sublimation	process by which a solid changes directly to a gas; occurs when ice or snow changes directly to water vapor (Lesson #68)
Subsoil	the bottom layer of soil which is made of large soil particles and some pieces of rock
Surface Water	water that is above ground in lakes, rivers, and oceans (Lesson #41)
Switch	a device that opens and closes an electric circuit (Lesson #101)

Help Pages

Glossary

Tectonic Plates	continental or oceanic plates that float over the surface of the Earth's mantle (Lesson #27)
Temperate Forest	a moderate climate ecosystem which is rich in plant and animal life
Temperature	a measure of average kinetic energy (Lesson #84)
Tertiary Consumer	third level consumer; an animal that eats animals that eat other animals (Lesson #12)
Thermal Energy	heat energy; related to the speed of the particles (Lesson #79)
Thermosphere	one of the outermost layers of the atmosphere (Lesson #50)
Tissue	a group of cells that work together to form organs (Lesson #16)
Topsoil	the top layer of soil; contains a mixture of various rock particles, air, water, and decayed organic material
Transfer of Energy	the movement of energy (Lesson #84)
Translucent	one of the properties of minerals; allows only a little light to pass through (Lesson #109)
Transparent	one of the properties of minerals; allows light to pass through (Lessons #37, 109)
Troposphere	the layer of atmosphere covering Earth's entire surface and containing 90% of all the gases in the atmosphere (Lesson #50)
Unbalanced Force	a force that causes a change in motion; a force that is not cancelled out by another force (Lesson #114)
Uninhabitable	not habitable; not able to support life
Vaporization	evaporation; process of changing from liquid to gas (Lesson #68)
Variable	any factor that can vary or change in an experiment (Lesson #9)
Velocity	speed in a specific direction (Lesson #119)
Verify	prove (Lesson #14)
Vertebrate	an animal that has a skull and a backbone (Lesson #19, chart)
Vibrations	back and forth movements (Lesson #107)
Volume (Matter)	the amount of space that matter takes up (Lesson #61)
Volume (Sound)	the loudness of a sound (Lesson #107)

Solutions© Science

Help Pages

Glossary

Warm Front	a warm moist air mass that rises and moves in to replace colder air; may bring rain (Lesson #57)
Waste Disposal	process by which an ecosystem disposes of its own waste by constantly recycling organic material (Lesson #18)
Water Cycle	the process by which liquid water continually evaporates, condenses, and falls to Earth as precipitation (Lesson #39)
Water Vapor	water that has evaporated; water in its gaseous state (Lesson #39)
Weathering	the wearing away of rock by water, wind, and ice (Lesson #27)
Wedge	simple machine with a slanted side and a sharp edge for cutting (chart)
Weight	a measure of the force of gravity on an object (Lesson #62)
Wheel-And-Axle	a wheel with a rod (called an axle) through its center; used to move heavy loads (chart)
Wind Sock / Wind Vane	a weather instrument which shows the direction of the wind (Lesson #54 & chart)
Work	effort or activity; what is done when a force is applied to an object over a distance
Zoologist	a scientist who specializes in the study of animals (Lesson #21)

Help Pages

Animal Groups

Invertebrates

Most of the members of the Animal Kingdom are invertebrates. An invertebrate is a multi-celled organism that does not have a backbone (vertebrae) or a bony inner skeleton. Some invertebrates do have a hard outer shell called an exoskeleton; others have only a soft body; still others have a fluid-filled skeleton. The chart shows some of the sub-groups of invertebrates.

Invertebrate	Illustration	Description	Examples
Annelid		segmented bodies; may be parasitic; prefer moist environment	earthworm, leech
Arthropod		segmented body; hard exoskeleton, jointed legs; multiple limbs	insect, spider, centipede, shrimp, scorpion, crayfish
Mollusk		soft body covered by hard shell; some live on land, others in ocean	snail, slug, squid, oyster, clam, cuttlefish, nautilus
Echinoderm		live in the oceans; spines and arms spread out from center of body	starfish, sea urchin, sand dollar, sea cucumber

Help Pages

Animal Groups

Vertebrates
Vertebrates are highly developed animals that have backbones and spinal chords. Only about 2% of all the animals in the world are vertebrates, but these are the animals we know best. That may be due to the fact that most vertebrates are much larger and take up more space than invertebrates. Also, vertebrates are very mobile – that means they can get around easily, and they tend to take control of the most favorable habitats.

Vertebrate	Illustration	Description	Examples
Amphibian		eggs hatch in water; young breathe with gills; adults develop lungs & live on land	salamander, frog, toad, newt
Bird		have beaks, wings, and feathered bodies; hollow bones for easy flight	crane, duck, robin, hawk, owl, penguin, ostrich, crow, swallow, bald eagle, chicken
Fish		most lay eggs; live in salt or fresh water; breathe with gills; use fins & tails to swim	salmon, shark, tuna, clownfish, marlin, baracuda, catfish, eel, perch, trout, blowfish, carp,
Mammal		give birth to fully developed young; hair or fur-covered bodies; feed young with milk	tiger, monkey, rat, seal, wolf, dolphin, whale, kangaroo, cat, raccoon, bear, squirrel, human
Reptile		breathe with lungs; may live on land or in water; bodies covered with scales	alligator, turtle, snake, gecko, iguana, crocodile, komodo dragon, chameleon

Help Pages

Animal Groups

Mammals
Most mammals have bodies that are covered with hair or fur. A mammal is warm-blooded which means it is able to regulate its body temperature. Most mammals give birth to fully formed babies, and mammal mothers produce milk to feed their young. Mammals are classified in many different ways. The chart shows some of the sub-groups of mammals.

Mammal	Illustration	Description	Examples
Cetacean		lives in water; equipped with tails & fins for swimming and blowholes for breathing	beluga, orca, blue whale, narwhal, humpback, dolphin, porpoise
Marsupial		babies not fully developed at birth; lives in mother's pouch during early development	kangaroo, wallaby, koala, wombat, Tasmanian devil, Virginia opossum
Carnivore		has four large canine teeth; highly developed brain; consumes animal flesh; some are omnivores	dog, bear, fox, raccoon, seal, walrus, tiger, weasel, skunk, lion, leopard, hyena, wolf
Primate		highly developed brain; arms, legs, hands with fingers & opposable thumbs	monkey, baboon, orangutan, chimpanzee, gorilla, human
Rodent		large incisor teeth; lives above ground & burrows underground; hibernates in winter	beaver, guinea pig, rat, porcupine, chipmunk, squirrel, gerbil, mouse, prairie dog

Help Pages

Animal Cell

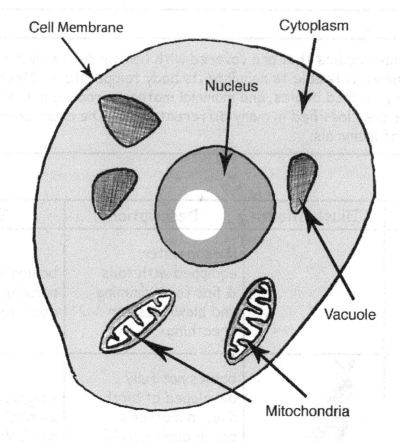

Plant Cell

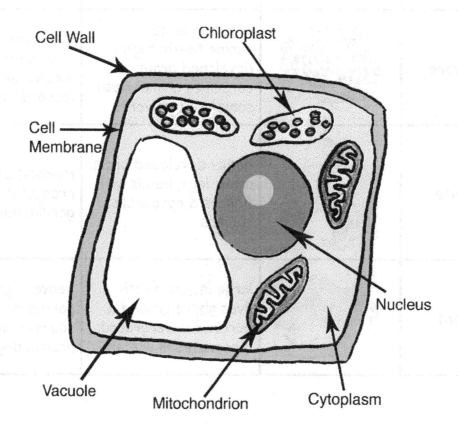

Help Pages

Clouds Types

Type	Description	Image
cirrus	Latin for "curl of hair"; wispy, featherlike, high altitude clouds	
stratus	Latin for "spread out"; low altitude horizontal sheets of clouds	
cumulus	Latin for "heap" or "pile"; dense, white, fluffy clouds	
cirrocumulus	series of small rippling cloudlets	
cumulonimbus	very dense, heavy storm clouds	
stratocumulus	low-lying, horizontal layers of clouds	

Help Pages

Cloud Terms

Latin Word	Signals	Cloud Name	Description
cirrus	curl	cirrus	wispy, like spider webs or feathers, high clouds
cumulus	heap or pile	cumulus	puffy, rippled or piled up, like cotton balls
stratus	spread out	stratus	layered, like blankets, mattresses, or waves
alto	middle (from Latin meaning "high")	altocumulus / altostratus	puffy & patchy / thin & uniform
nimbus	rain (from Latin meaning "cloud")	cumulonimbus / nimbostratus	storm clouds / dark, low layers

Earth

Earth's Layers

The four layers of Earth are crust, mantle, outer core, and inner core. The crust is the thinnest layer, and it varies in thickness because of landforms. The crust may be 5 km thick in some places and as much as 70 km thick in other places. The mantle is the thickest layer; it's about 2900 km thick. The solid upper part of the mantle and the crust make up the lithosphere – Earth's outer shell. The outer core is made of a liquid – molten iron. The inner core is solid but it is the hottest layer. In fact the center of the Earth is nearly as hot as the sun.

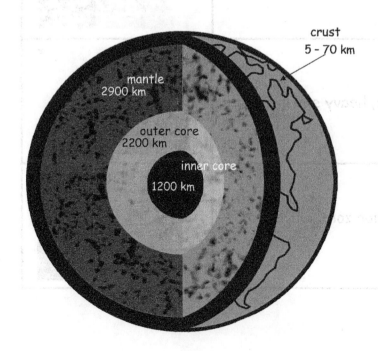

Although the Earth's crust is its thinnest layer, it contains many layers of soil and subsoil made of organic matter, sediments, sand, minerals, and rock. Beneath all of these layers is **bedrock**.

Simple Solutions© Science Level 5

Help Pages

Metamorphosis

Incomplete Metamorphosis

Incomplete Metamorphosis involves three stages of development: egg, nymph, and adult. The nymph looks very similar to the adult. The nymph and the adult have the same habitat and diet. Insects that go through incomplete metamorphosis include cockroaches, grasshoppers, and dragonflies.

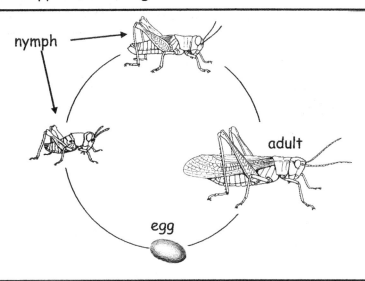

Complete Metamorphosis

Complete Metamorphosis involves four stages: egg, larva, pupa, and adult. The insect starts out as an egg. When it hatches, it is in the larva stage. Caterpillars are butterfly larva. Larvae eat as much as they can. Then they go into a dormant (inactive) state called the pupa. Butterflies spend the pupa phase in a cocoon. When the insect comes out of the pupa, it is a fully grown adult. Beetles, moths, and flies are some of the other insects that go through complete metamorphosis.

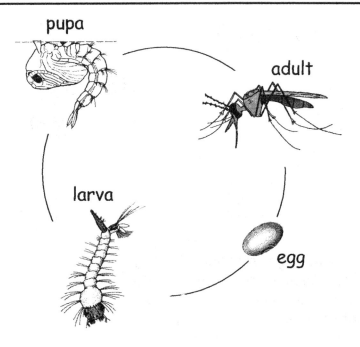

Help Pages

Metamorphosis

Metamorphosis of Amphibians

Amphibians like frogs, toads, salamanders, and newts also go through a **metamorphosis**. These cold-blooded animals are hatched from eggs and must live in water during the first phases of life. Newborn amphibians have no legs, and they breathe through gills. They change shape as they grow, developing lungs and legs while losing their gills and tails. As adults, amphibians live on land but always prefer to be near water.

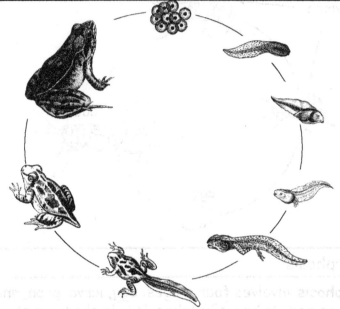

Organism — Environmental Interaction

Primary consumers, like rabbits, eat plants to get energy to live and grow.

Plants use energy from the sun, along with water and nutrients from the soil, to make food during photosynthesis.

Secondary consumers, like foxes, eat animals (rabbits) to get energy to live and grow.

Decomposers break down dead organisms like the fox's body after it dies. Decomposers use the energy from dead plants and animals. They also put nutrients back into the soil for plants.

Help Pages

Simple Machines

Machine	Description	Example	Image
pulley	uses grooved wheels and ropes to raise and lower things	flagpole lift, clothesline, window blinds	
lever	bar that pivots on a fulcrum to lift or move heavy loads	seesaw, shovel, crowbar	
wedge	has a slanted side and a sharp edge for sliding or for cutting	ax, knife blade, garden hoe, scissors	
wheel-and-axle	wheel with a rod (axle) through its center used to move loads	wheel, doorknob, steering wheel	
inclined plane	a slanted surface (also called a ramp); used to move things to higher or lower places	boat ramp, wheelchair ramp, sliding board	
screw	an inclined plane spiraled around a post used to fasten or hold things together	light bulb neck, screw-top on a bottle, spiral staircase	

Simple Solutions© Science Level 5

Help Pages

Laboratory Instruments

Tool	Measures	Units	Image
Balance	measures an object's mass (the amount of matter in the object)	grams kilograms	
Thermometer	used to measure temperature	degrees	
Spring Scale	used to measure forces like weight and friction	Newtons	
Beaker	holds and measures the volume of liquids	liters milliliters	
Ruler or Measuring Tape	measures the length and width of objects	meters centimeters inches	

Help Pages

Laboratory Instruments (continued)

Tool	Use	Image
magnifying glass (hand lens)	magnify	
microscope	magnify the view of tiny objects to hundreds of times their natural size	
dropper / pipette	measure out small amounts of liquid	
forceps	hold or pick up small objects	
safety goggles	provide eye protection	

Simple Solutions® Science

Help Pages

Laboratory Instruments (continued)

Tool	Use	Image
magnifying glass (hand lens)	enlarges	
microscope	magnifies the view of tiny objects to hundreds of times their normal size	
dropper / pipette	measures out small amounts of liquid	
forceps	to pick up small objects	
safety goggles	provide eye protection	